Keynotes

Requirements Engineering in the Age of Disruptive Digital Transformation

Eric Yu

University of Toronto, Canada

Abstract. Recent emerging technologies such as social, mobile, cloud and particularly data analytics are rapidly transforming how organizations operate, leading them to rethink their business models and strategic positioning. What requirements engineering techniques are needed in this age of rapid and frequently disruptive change? Goal modeling with strategic actor relationships, such as in i*, could be a starting point. In this talk, I will outline some tentative steps towards a modeling framework.

Urban Computing: Using Big Data to Solve Urban Challenges

Yu Zheng

Microsoft Research, China

Abstract. Urban computing is a process of acquisition, integration, and analysis of big and heterogeneous data generated by a diversity of sources in cities to tackle urban challenges, e.g. air pollution, energy consumption and traffic congestion. Urban computing connects unobtrusive and ubiquitous sensing technologies, advanced data management and analytics models, and novel visualization methods, to create win-win-win solutions that improve urban environment, human life quality, and city operation systems. Urban computing is an inter-disciplinary field where computer science meets urban planning, transportation, economy, the environment, sociology, and energy, etc., in the context of urban spaces. In this talk, I will present our recent progress in urban computing, introducing the key applications and technologies for integrating and deep mining big data. Examples include fine-grained air quality inference throughout a city, city-wide estimation of gas consumption and vehicle emissions, and diagnosing urban noises with big data. The research has been published at prestigious conferences (such as KDD and UbiComp) and deployed in the real world. More details can be found on http://research.microsoft.com/en-us/projects/urbancomputing/default.aspx

Contents

Requirements Engineering Tools

Automated Requirements Analysis

Automated Service Selection Using Natural Language Processing

Muneera Bano[1]([⊠]), Alessio Ferrari[2], Didar Zowghi[1], Vincenzo Gervasi[3], and Stefania Gnesi[2]

[1] University of Technology Sydney, Ultimo, Australia
{Muneera.Bano,Didar.Zowghi}@uts.edu.au
[2] CNR-ISTI, Pisa, Italy
{alessio.ferrari,stefania.gnesi}@isti.cnr.it
[3] Dipartimento di Informatica, Università di Pisa, Pisa, Italy
vincenzo.gervasi@di.unipi.it

Abstract. With the huge number of services that are available online, requirements analysts face an overload of choice when they have to select the most suitable service that satisfies a set of customer requirements. Both service descriptions and requirements are often expressed in natural language (NL), and natural language processing (NLP) tools that can match requirements and service descriptions, while filtering out irrelevant options, might alleviate the problem of choice overload faced by analysts. In this paper, we propose a NLP approach based on *Knowledge Graphs* that automates the process of service selection by *ranking* the service descriptions depending on their NL similarity with the requirements. To evaluate the approach, we have performed an experiment with 28 customer requirements and 91 service descriptions, previously ranked by a human assessor. We selected the top-15 services, which were ranked with the proposed approach, and found 53% similar results with respect to top-15 services of the manual ranking. The same task, performed with the traditional cosine similarity ranking, produces only 13% similar results. The outcomes of our experiment are promising, and new insights have also emerged for further improvement of the proposed technique.

Keywords: Service selection · Requirements engineering · Knowledge graphs · Natural language processing

1 Introduction

Software as a service (SaaS) offers reusability and flexibility while reducing time and costs for the developers [16]. This style of development has introduced new challenges, especially for analysts, when they have to find the best matching service for the requirements of all concerned stakeholders, while making a trade-off between cost, functional and quality requirements [5]. Identifying and selecting the adequate service is the most crucial step in service oriented software development [12]. It is considered a challenging task due to various reasons, e.g.,

© Springer-Verlag Berlin Heidelberg 2015
L. Liu and M. Aoyama (Eds.): APRES 2015, CCIS 558, pp. 3–17, 2015.
DOI: 10.1007/978-3-662-48634-4_1

mismatch in the level of abstraction and granularity of customer requirements and service descriptions, lack of context in services, and, in particular, overload of choice [18]), due to the large number of available services with similar characteristics [5]. Indeed, over the last decade, the number of third party services provided over the internet has increased substantially. This has exacerbated the situation for analysts when they have a huge set of services to evaluate, and select one that best serves the customer's requirements [3]. Various solutions have been proposed (e.g., [12, 23]) that employ a variety of techniques for service identification, such as guidelines, patterns, business process models, and value analysis. However these approaches require certain degree of formalism to identify the adequate service for the requirements, and work only with a small and manageable number of services.

In this paper, we present a natural language processing (NLP) approach that uses Knowledge Graphs (KG) [9, 10] for dealing with the problem of overchoice [4]. The proposed approach has been evaluated in an experiment with a realistically large number of service descriptions – i.e., 91 descriptions retrieved from the Web. The approach was compared with a manual evaluation, and with a more traditional information retrieval approach based on cosine similarity [19]. The KG-based approach substantially outperforms the one based on cosine similarity in selecting the most relevant services, after filtering out irrelevant options.

The major contributions of this research are twofold: (a) the use of a novel NLP method and supporting tool for substantially decreasing the number of available services found by search engines that match a set of requirements, and (b) a methodology to support requirements analysts in reducing the time and effort needed for optimal service selection.

This paper is organized as follows. Section. 2 provides an overview of the service selection task to be addressed, as well as the automated techniques employed in this paper. Section. 3 presents the experimental evaluation performed and Section. 4 discusses the results achieved, while Section. 5 provides comments concerning the threats to validity that might have affected our work. Section. 6 provide pointers to related works. Finally, Section. 7 concludes the paper.

2   Computing Similarity

To address the problem of "choice overload", our goal is to automatically discard those service options that are less relevant for the requirements, and present a reduced list of services, which can be browsed by the analyst to perform a selection within a more manageable set of choices. To this end, we wish to provide an automated technique that *ranks* the services according to their degree of satisfaction with respect to the set of requirements. Once a ranked list of services has been automatically produced, the analyst can discard the services that have lower ranking. The idea of the approach is that automated ranking can be performed by computing the *similarity* among the NL content of the requirements and the NL descriptions of the services. The underlying assumption is that if a requirement and a description of a service are semantically similar in terms of NL content, the service is likely to satisfy the requirement.

More formally, given a set of requirements $R = \{r_0, \ldots, r_m\}$, and a set of descriptions of services $S = \{s_0, \ldots, s_l\}$, we wish to rank the services according to their degree of matching with respect to the requirements. To this end, we compute the similarity σ of a requirement r_i with respect to a service description s_j. The overall ranking K of a service description s_j with respect to a set of requirements R is given by:

$$K(s_j, R) = \sum_{i=0}^{m} \sigma(r_i, s_j).$$

Given $|S|$ service descriptions, we can compute the ranking K of each service, and order the services according to K. Then, the analyst can discard the k services that have lower ranking, and manually analyse the $|S| - k$ services with higher ranking. In this paper, $\sigma(r_i, s_j)$ will be implemented by two functions, namely the Knowledge Graph similarity function, and the classic cosine similarity [19] function, which are described in the following sections.

2.1 Knowledge Graph Similarity

A Knowledge Graph [10] is a representation of a document set D in the form of a directed weighted graph $KG = \{V, E, w\}$, where V are nodes, E are edges and w is a weighting function. In our context, the set of documents D is composed of those documents that give the definitions of the concepts included in the requirements. For example, if – as in our experimental evaluation (Sect. 3) – the requirements concern an SMS gateway service, the documents in D will include a NL definition of what a SMS gateway is, a description of the standard protocols considered in the requirements (e.g., HTTP, SOAP), as well as other documents describing relevant concepts mentioned in the requirements. For a precise definition of E, V and w, the reader should refer to [10]. Here, we give the essential information required to understand the rest of the paper.

Nodes. Each node $v \in V$ in the KG is labeled with the morphological root of a relevant term contained in D. The relevant terms are those terms that are not part of the so-called *stopwords*, i.e., words which are common in English such as pronouns, prepositions, articles and conjunctions. The use of morphological roots (also called *stems*) implies that terms such as "communication" and "communicator" are both represented by the same node "communic". We call $\overline{T}$ the stems of the unique terms in D that are not stopwords.

Edges. Each directed edge in the graph is associated with the *co-occurrence* of two terms in the document set. We say that two terms co-occur when they appear *aside* each other in a sentence. Co-occurrence normally indicates semantic proximity among concepts [20], and therefore we use it to represent concepts relationships. The nodes of our graph are not labeled with terms, but with stems. Hence, the concept of co-occurrence is in this case related to stems. The edges $e = (v_i, v_j) \in E$ in our graph are all those couples of nodes such that the corresponding labels of the nodes co-occur in D. The direction of each edge

represents the order in which the stems connected by the edge can appear in a
sentence of the document set.

Weighting Function. Each edge is labeled with a real positive number according to the weighting function w. The weighting function $w : E \rightarrow [0,1]$ is the
inverse function of the semantic relatedness of two stems (i.e., 1 divided by the
number of their co-occurrences). The more two stems connected by an edge
co-occur in D, the lower the value returned by w.

An example excerpt of a KG extracted from the Wikipedia document for
the term "SMS" is presented in Fig. 1. We see that the weight of the edges is
lower for those stems that frequently occur together in the document (e.g., for
$e = $ ('text', 'messag'), $w(e) = 0.058$), and have a tighter semantic connection
in the domain. Higher weights (e.g., $e = $ ('protocol', 'SMS'), $w(e) = 0.5$) or no
edge (e.g, between 'mobil' and 'short') occur when the semantic connection in
the domain is weaker.

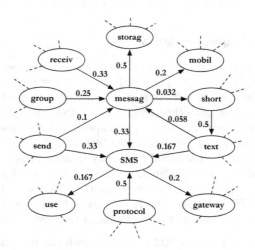

Fig. 1. Excerpt of a Knowledge Graph

KG-Expansion. Given a term, we wish to expand it with the set of concepts
associated to such term in the KG. To this end, we define the term-expansion
function $I : \overline{T} \times KG(D) \rightarrow 2^{\overline{T}}$ as the function that associates the stem of a
term to all the stems that are directly connected to such stem in the KG. More
formally, given a stem $\overline{t}$ in $\overline{T}$, we define the term-expansion function as:

$$I(\overline{t}, KG(D)) = \{\overline{s} \in \overline{T} : \exists e \in E, e \in \{(\overline{t}, \overline{s}), (\overline{s}, \overline{t})\}, w(e) \le \epsilon\} \cup \overline{t}$$

The value $\epsilon \in [0,1]$ is used to discard the stems connected with low semantic
proximity – according to our experience, we suggest to choose $\epsilon \le 0.5$. If we
apply such function to the term "SMS", we consider the KG in Fig. 1, and we

set $\epsilon = 0.4$ such function will return the set ('SMS', 'messag', 'text', 'send', 'use', 'gateway'). The term-expansion function can easily be extended to the sets of non-stopwords stems $\bar{q} = \{\bar{q}_1, \ldots, \bar{q}_{|\bar{q}|}\}$ of a generic input document q. We call this function KG-expansion function $\mathcal{I}$, defined as:

$$\mathcal{I}(\bar{q}, KG(D)) = \bigcup_{j=1}^{|\bar{q}|} I(\bar{q}_j, KG(D))$$

Given a requirement or the description of a service – which can be both regarded as documents q – these can be represented through the sets of their stems. Stopword removal can be applied to represent them as $\bar{q}$. Afterwards, such sets can be expanded through the KG-expansion function, which basically extends the concepts included in the service description or requirement with the associated concepts in the KG.

KG-Similarity. The KG-similarity function aims to compute the similarity between a requirement r_i and a service description s_j with the support of the KG. Such similarity function is a Jaccard similarity metric [19] based on the expansion of both the requirement and the service description through the function $\mathcal{I}$. We define the KG-similarity function between r_i and s_j as:

$$\sigma_{KG}(r_i, s_j) = \frac{|\mathcal{I}(\bar{r}_i, KG(D)) \cap \mathcal{I}(\bar{s}_j, KG(D))|}{|\mathcal{I}(\bar{r}_i, KG(D)) \cup \mathcal{I}(\bar{s}_j, KG(D))|}$$

In a sense, this similarity metric measures the proportion of stems that the extended version of a requirement and the extended version of a service description have in common, and it assumes that the greater is the number of shared stems, the higher is the chance of the requirement to be satisfied by the service description. The approach helps dealing with issues related to synonyms in requirements and service descriptions. Since synonyms tend to occur in similar linguistic contexts – i.e., they co-occur with similar terms –, we expect the KG-expansions of synonyms to have several stems in common. In turn, we expect the KG-similarity function to detect this high degree of *contextual* similarity.

2.2   Cosine Similarity

Vector Space Model. In information retrieval [15], it is typical to represent documents according to the vector-space model [17], which has been also employed in requirements engineering to support the computation of the similarity among requirements [6,8]. With this model, a natural language document q is regarded as a sparse vector $q = \{q_{u_0}, \ldots, q_{u_h}\}$, where each vector component q_u is associated to a term u in the vocabulary of the vector space. Such vocabulary is made of all the terms included in the documents to be evaluated. In our case, the vocabulary is made of all the terms in the service descriptions.

The value of a component q_u is 0 if the term u does not appear in q, and is included in $(0, 1]$ if the term appears in q. The value of q_u scores the relevance of the term for the document q. Such relevance is normally computed by means of the $tf - idf$ score [15].

COS-Similarity. The cosine similarity function aims to compute the similarity between a requirement r_i and a service description s_j within the vector space, whose vocabulary is made of all the terms included in the service descriptions. Each r_i and each s_j are first represented according to the vector space model as $\boldsymbol{r}_i$ and $\boldsymbol{s}_j$. Then, their similarity is evaluated according to the cosine similarity function σ_{COS} defined as:

$$\sigma_{COS}(r_i, s_j) = \frac{\boldsymbol{r}_i \cdot \boldsymbol{s}_j}{|\boldsymbol{r}_i||\boldsymbol{s}_j|}$$

Such similarity measures the cosine of the angle between the two vectors, and it assumes that the more relevant terms the requirement and the service description have in common, the higher is the chance of the requirement to be satisfied by the service description. In our implementation we have discarded stopwords and we have employed *stems* instead of terms in our vector space model. Therefore, in the definitions given above, each term u shall be considered a non-stopword stem $\overline{u}$.

Table 1. Excerpt of the requirements set adopted in our experiments (top), and excerpts of two service descriptions (bottom).

ID	Requirement
0	Service supports outgoing text messages in Australia
1	Service should not have any hardware or SIM requirements
2	Service should be highly reliable with 99.9% message delivery
...	...
22	SMS transmission delay should be between 3 to 7 seconds
23	Service should offer contacts management
24	Service supports retrieval of send messages record
25	Service shows message delivery message
26	Service shows message delivery failure notification
27	Service should provide schedule message delivery

Messente	Intel Tech
Send SMS online, SMS gateway API	Why Choose Us?
Easiest group SMS messaging in the universe	Cloud Services
Start sending SMS to your customers anywhere	Lightening Fast Delivery Times
in the world within 60 seconds	Amazing 24/7 Support
Send SMS to 200+ countries	Reliable 100%
No fees	Uptime SLA
...	...

3   Experimental Evaluation

The data used for the experiment was about an SMS gateway service selection, which enables Websites to send and receive text or multimedia messages with simple invocation of the remote service API while hiding all the underlying technical and infrastructure details. The same data was previously used in a case study on service selection [4]. Online searches resulted in 91 eligible options providing the SMS gateway services, which were to be evaluated against the 28

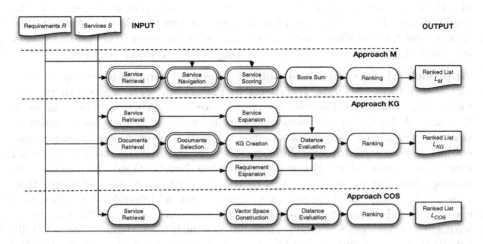

Fig. 2. Overview of the different treatments applied. Double-lined ellipses indicate manual activities.

requirements in [4] (an excerpt is shown in Table 1 - top). The list of 91 services along with the links to their descriptions are available online at http://goo.gl/CcguZM. Two representative excerpts of NL service descriptions (according to their Websites at the time of the experiments) are reported at the bottom of Table 1.

The goal of this experiment was to evaluate to which extent the KG-similarity function and the COS-similarity function could be employed to discard the less relevant services, and therefore address the problem of overload of choice by reducing the list of service options that a requirement engineer has to evaluate.

3.1 Design and Execution

The experiment was carried out separately by the first and second authors, referred in the following as subject 1 and subject 2, respectively. Fig. 2 shows the different approaches that have been applied in the experiment. The input of the experiment – equivalent for both subjects – was composed of the set of NL requirements $R = \{r_0, \ldots, r_{27}\}$, and the links to the main Web-page of the 91 services. The NL content of these Web-pages represented our service descriptions $S = \{s_0, \ldots, s_{90}\}$.

Subject 1 performed the service ranking task manually (approach M), and subject 2 performed the ranking task by first applying the KG-similarity function (approach KG), and then the COS-similarity function (approach COS). Approach KG includes a manual set-up approach, where the domain documents D have been retrieved and selected. Approach COS does not include a manual set-up and it is fully automatic. For simplicity, we refer to both approaches as "automated approaches". The result of each approach was an ordered list of services, ranked according to their degree of satisfaction with respect to the requirements. The ranking provided by means of approach M has been regarded as a *ground-truth* against which the automated approaches (KG and COS) have been tested in terms of ability in discarding the less relevant services.

Approach M. In approach M, subject 1 browsed through all the online service descriptions (**Service Retrieval** in Fig. 2) and gave a score to the services against each requirement (**Service Scoring**). Whenever the content of the main Web-page of the service description was not giving enough information to evaluate the degree of satisfaction with respect to the requirements, subject 1 navigated the Web-links provided in the main Web-page to gather more information about the service (**Service Navigation**). In a sense, the strategy followed by subject 1 was to use the content of the Web-pages (i.e., $S = \{s_0, \ldots, s_{90}\}$), and extend such content with additional information that could be found within the Web-links. For each requirement, a service was scored with three values i.e. 0= requirement not satisfied; 0.5= requirement partially satisified; 1= requirement completely satisfied. Then, the scores obtained by the service on each requirement were summed up to obtain a ranking score K for the service (**Score Sum**).The output of approach M was a ranked list of services L_M (**Ranking**), with associated ranking score K computed according to the scoring schema described above. The overall time required to perform the evaluation with approach M was two weeks.

Approach KG. In approach KG, subject 2 automatically retrieved the NL content of the Web-pages of the services to compare such content (i.e., $S = \{s_0, \ldots, s_{90}\}$) with the requirements (**Service Retrieval**). Then, subject 2 selected a set of documents D to build the knowledge graph. Such set was selected by automatically downloading the textual content of the Wikipedia pages that were associated to the terms included in the requirements (**Documents Retrieval**). More specifically, for each term in the requirements, the Wikipedia page associated to that term was downloaded, when available. The downloaded documents were briefly reviewed to discard irrelevant pages (e.g., the Wikipedia page for "soap" was referring to the cleaning soap and not to the SOAP protocol). Additional domain documents were included in D for those technical terms that were considered more relevant, namely "HTTP", "PHP", "SMS gateway service", "SMS", "SOAP". In total, D included 28 documents, 23 automatically selected and 5 manually added. These two activities are referred in Fig. 2 with the task named **Documents Selection**. Finally, subject 2 performed the experiments with a Python implementation of the approach described in Sect. 2.1, which consists in building a knowledge graph from the documents (**KG Creation**), and then expanding requirements and services by means of the knowledge graph (**Requirement Expansion** and **Service Expansion**, respectively). The KG extracted from the documents resulted in $|V| = 6683$ nodes (i.e., individual stems) and $|E| = 37253$ edges (i.e., co-occurrences). Evaluation of the KG-similarity (**Distance Evaluation**) has been performed with $\epsilon = 0.5$, hence all the edges with weight 1 – i.e., associated to single co-occurrences – have been discarded. The output of the experiment was a ranked list of services L_{KG} (**Ranking**), with associated ranking score K computed by summing-up the contributions of the KG-similarity function on each requirement. The overall time required by approach KG was 25 minutes.

Approach COS. In approach COS, subject 2 used the service descriptions already retrieved with approach KG (**Service Retrieval**). Then, subject 2 performed the ranking task with a Python implementation of the approach described in Sect. 2.2, which consists in building the vector space (**Vector Space Construction**) and in evaluating the distance between services and requirements through the cosine similarity metric (**Distance Evaluation**). The output of the experiment was a ranked list of services L_{COS} (**Ranking**), with associated ranking score K computed by summing-up the contributions of the COS-similarity function. The overall time required by approach COS was 2 minutes.

3.2   Results

To evaluate the results of the experiments, we measured the "degree of correspondence" between the ranking of the *ground-truth* (approach M) and the rankings of approach KG and COS, in terms of services that have been considered less relevant – and therefore discarded – by the three approaches. The index adopted to evaluate such "degree of correspondence" is the accuracy, which, in our context, will be referred as *filtering accuracy*. Given the ranked list L_M of approach M, and given the ranked list L_A of one of the automated approaches, let O_M be the set of services discarded by approach M, and let O_A the set of services discarded by one of the automated approaches. Moreover, let k be the number of services discarded in both approaches (i.e., $k = |O_A| = |O_M|$). The filtering accuracy is:

$$\alpha = \frac{|O_M \cap O_A|}{k}.$$

The value of α has been computed by varying k in $[1, |S|]$, with $|S| = 91$, for our experiments. The idea of such evaluation is: if the requirements engineer discards the k less relevant services form L_A, how many of the services discarded belong to the k less relevant services of L_M? Since L_M comes from a human evaluation, high values of α gives confidence on the fact that the automated approach discards the same services that would be discarded by a human.

Moreover, we have also computed an index of accuracy that considers how many relevant services are retrieved by the automated approaches. We call this index *selection accuracy*, and we define the index as follows (Also in this case, we evaluate ρ by varying k in $[1, |S|]$):

$$\rho = \frac{|(L_M \setminus O_M) \cap (L_A \setminus O_A)|}{|S| - k}.$$

Fig 3 plots the filtering accuracy for approach KG and approach COS (α_{KG} and α_{COS}, respectively), for increasing values of k. Let us now suppose that the

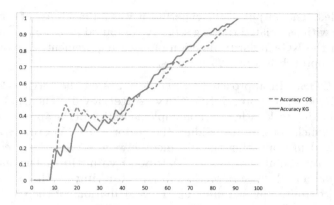

Fig. 3. Results for the filtering accuracy.

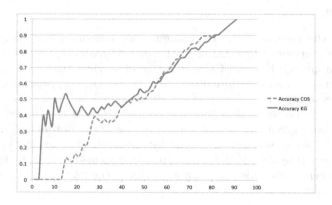

Fig. 4. Results for the selection accuracy.

analyst wishes to filter out the 90% of the irrelevant services (top-right part of Fig. 3). We see that, to have $\alpha_{KG} > 0.9$ (i.e., a 90% filtering accuracy), we must have $k > 75$. This implies that the analyst can focus on the top-16 services, being confident that 90% of the irrelevant services have been correctly filtered. Slightly lower performances are achieved for α_{COS}. Indeed, the filtering accuracy is above 0.9 only for $k > 82$, which implies that the analyst has to discard 82 services from the ranked list, to be sure to filter out the 90% of irrelevant services. However, as visible by looking at the top-right part of Fig. 3, the ability of filtering the irrelevant services for high values of k can be considered comparable for the two methods. It is also worth noting that, for low values of k, the filtering accuracy of approach COS is considerably higher than that of approach KG (see left part of Fig. 3). We argue that this might be due to an *inclusive* tendency of the KG-expansion function. Indeed, such function expands the concepts of a service description with additional concepts, and even though the description is rather poor in terms of requirement coverage, it could become richer thanks to the additional concepts included in its KG-based expansion. However, such inclusive nature appears to have a negative effect only for lower values of k. For high values of k (i.e., if we consider only the top-ranked services), the effect of such inclusive nature appears to be negligible with respect to the positive effect in providing contextual concepts, which seems to provide a more accurate requirement-to-service matching. Let us now consider the selection accuracy. Fig 4 plots the selection accuracy for approach KG and approach COS (ρ_{KG} and ρ_{COS}, respectively), for increasing values of $|S| - k$. From the plot, we see that, given a set of 15 top-ranked services (local maximum on the left part of the graph), approach KG is able to identify 53% of the relevant services, while approach COS identifies only the 13% of the relevant services. To assess the effectiveness of the approaches, it is useful to compare the selection accuracy on the top-15 results with a random predictor model. Such a model assumes to randomly select $k = 15$ items among the $|S| = 91$ items: the accuracy for such a model is $k/|S| = 0.16$. Hence, the selection accuracy is 16% for the random predictor, 53% for approach KG and 13% for approach COS. Therefore, approach KG outperforms the random predictor by 37%, while for approach COS the accuracy is even lower by 3% with respect to the random predictor. Note that having a local maximum of accuracy for $k = 15$, is an encouraging result in practical terms. Indeed, given a ranked list of results, as occurs for search engines, people tend to focus on the first page of the results (see, e.g., https:// chitika.com/google-positioning-value), which normally displays 10 to 15 items. We conjecture that, for an analyst, 15 items can be considered a manageable set also in the case of service selection. To have a clear view of this result, it is useful to look at the tables of the top-15 services of the three approaches.

Table 2 (left) reports the top 15 services together with their rank according to the manual assessment, while Table 2 (center) reports the top 15 services obtained with approach KG. The 8 services in common in the two lists are highlighted in **bold**. The tangible results presented in Table 2, can give us confidence on the effectiveness of the method: if we consider a requirements analyst who searches for the best service that satisfy a set of requirements, he/she can have confidence that the top results of the approach will be likely to be suitable for its needs. Then, he/she can manually review the top results to select the best service.

Instead, the top-15 results obtained with approach COS (Table 2 (right)) show that only two of these results appear in Table 2 (left). Therefore, in a practical scenario, the analyst would have gone through the top-15 services, and found that really few of them were in line with his/her requirements.

4   Discussion and Improvements

The major contribution of the study is the automation of the process of matching requirements against a big set of service descriptions by means of the KG approach. The KG approach has yielded 53% selection accuracy, and 90% filtering accuracy within a manageable reduced list of services. We argue that

Table 2. Ranking for approach M (left), KG (center), and COS (right).

Service	K
Intel Tech	27.5
Skebby	27.5
Direct SMS	27
Via SMS	27
Messente	26.5
Ready to SMS	26
Click Send	25
Bulk SMS	24.5
Clock Work SMS	24.5
Red Oxygen	24.5
Clickatell	24
SMS Broadcast	24
SMS Global	23.5
Essendex	23.5
Nexmo	23.5

Service	K
Messente	7.58
One API 4 SMS	6.25
Red Oxygen	5.84
Intel Tech	5.82
Click Send	5.75
Wave Cell	5.74
Cdyne	5.62
Budget SMS	5.61
Plivo	5.47
Clock Work SMS	5.46
SMS Global	5.35
M4U	5.25
TXT Nation	5.25
Nexmo	5.16
Bulk SMS	5.09

Service	K
Text Impact	2.45
Text Anywhere	2.42
SMS Roaming	2.35
Vodafone Multitext SMS	2.33
Developer Garden	2.31
Oventus	2.20
Free SMS Craze	2.19
SMS Country	2.15
Text Marks	2.10
Carousel SMS	2.07
Essendex	1.98
Text Local	1.91
Ausie SMS	1.89
Ready to SMS	1.82
GSMA	1.68

the accuracy of the results may have been affected by the procedure for service browsing. The KG approach only used the first page of the Websites for service description to match them against the requirements. Whereas in manual evaluation, the researcher went through further to seek more information to score the services (i.e., a **Service Navigation** task was introduced). A similar technique can be implemented in the KG approach to traverse the relevant links on the first page to retrieve further information.

Another issue to be addressed in the KG approach is associated to its coarse-grained similarity computation. Indeed, requirements and services might include relevant fine-grained constraints. Consider the excerpt of the "Messente" service description in Table 1, with respect to requirement 22: *"SMS transmission delay should be between 3 to 7 seconds"*. From the description, which tells *"Start sending SMS to your customers anywhere in the world within 60 seconds"*, we see that the service is able to send messages within 60 seconds. Therefore, it is likely to fail in satisfying requirement 22. Nevertheless, the "Messente" service is evaluated as the top-service for the KG approach. This result is due to two factors: (1) summing-up the values of the contributions of the different requirements to provide an overall ranking might hide those requirements for which the service is not the best; (2) our measure of similarity among a requirement and a description focuses on the "topic" of a requirement (e.g., the transmission delay), and does not consider the lower-level semantic dimension (e.g., the actual temporal constraint). The first issue can be easily addressed, by considering the similarity of the *single* requirement (i.e., the KG-similarity) with respect to the whole service, and avoiding to sum-up the contributions of each requirement. Therefore, the approach already include a way to spot out *outliers*, i.e., requirements not fully satisfied. Instead, to address the second issue, and achieve a finer grain similarity measure, specific heuristics have to be defined and integrated in the methodology.

A final aspect to discuss is the time required by the KG approach. Though 25 minutes are acceptable, it is still a rather high amount of time for an automated approach. The bottleneck of the approach is the **Service Expansion** task, which, in our experiments, required 17 minutes. Besides code-level optimization, given the higher efficiency of the COS approach, which required only 2 minutes in total, we argue that the two approaches could be combined as follows. The COS approach could be used for lower values of k, where its filtering accuracy is even higher than the KG approach. Then, the KG approach can be employed only on the pages that received higher ranking according to the COS approach. The computation of the exact threshold for the usage of one approach and the other (i.e., from which threshold value of k is preferable to use the KG approach) requires further studies with multiple data-sets.

5   Threats to Validity

In our experiments, we have used two treatments, namely approach KG and approach COS, to compare to a manual service selection treatment (approach M), considered as a reference baseline. As shown in Fig. 2, the input of each treatment was the same (i.e., requirements and services), and the treatments were evaluated according to a comparable output (i.e., a ranked list). Two performance measures have been employed to evaluate the results, namely the filtering accuracy and the selection accuracy. Moreover, we have also shown a human-understandable view of the results in the form of ranking tables. Therefore the design of the experiment provides confidence on the results achieved by approach KG. Concerning a possible researcher's bias, it is worth highlighting that the two researchers (subject 1 and 2) operated in parallel in two different institutions, and produced their evaluations independently.

The confounding variables that can affect the internal validity of the results could be attributed to the difference in the scoring procedure adopted by the manual and automated approaches. Furthermore, the retrieval technique used in the manual experiment was more comprehensive than that of the automated ones because, as we mentioned, more Web-pages were traversed to retrieve relevant information and has also impacted the results of the experiment. Though these two confounding variables could in principle threat the internal validity of our results, it is worth noting that, in the evaluation, the approaches were compared according to their input and output, which, by experimental design, are comparable. Hence, the effect of such variables is mitigated by the design of the experiment. The experiment design, data collection and execution are described in sufficient details to make it repeatable. However, the service descriptions that were retrieved from the online URL may change over time (e.g. in case of update or new release of the Web-site). Therefore, in that scenario, the replication with similar techniques may provide different results.

6   Related Works

Over the last decade, a significant amount of research has been conducted in providing tools, techniques and methods to deal with the task of service selection [14,16], and various researchers have proposed to *automate* the service identification process. Among these works, the SeCSE project [22] has the aim to create free and open source methods, tools and techniques for system integrators and service providers. Their main objective is to provide a mechanism where discovered services can be used for improving and refining the initial requirements and thus bringing better alignment of requirements and business needs to the services and vice versa. IBM has proposed the SOMA [2] approach, which is an end-to-end software development method for service oriented solutions. The SENSORIA [21] project had the aim to develop a new approach for service oriented software engineering with existing theories, techniques and methods for ensuring correctness of the procedure and allowing a semi-automatic design process. The usage of *business process* (BP) models for service identification is another relevant line of research. Jamshidi *et al.* [13] focus on the service-oriented development with BP models, while Adam *et al.* [1] propose to use BPs for identifying the services at the right level of abstraction and granularity. These approaches make the assumption that the BP models will have complete details of their corresponding processes, which is hard to achieve. These existing approaches require a certain degree of formalism (e.g., XML, UML, BP models) for comparing the requirements and service specifications and do not operate with specifications in NL and with overload of choice. With a considerably huge number of available services, conversion of service specifications to formal notations is a time consuming task. Our proposed KG approach reduces the sample of available services to a manageable size without effort of formalisation, and within a short time span.

7   Conclusion

With increase in the number of services offering similar functionality, analysts face the problem of *overchoice* when they have to select one service against requirements. In this paper we have presented an approach and supporting tool that automates the process of matching requirements against service descriptions using *Knowledge Graphs* [9,10]. Though the approach is promising to address the challenges of service selection, there is still room for improvement in the efficiency of the approach and accuracy of the results (see Sect. 4). Our future research will focus also on comparing the performance of the KG-similarity metric with the other existing approaches for computing NL requirements similarity (see, e.g., [7,11]), to assess their effectiveness in the field of service selection.

References

1. Adam, S., Riegel, N., Doerr, J.: Deriving software services from business processes of representative customer organizations. In: SOCCER 2008, pp. 38–45. IEEE (2008)
2. Arsanjani, A., Ghosh, S., Allam, A., Abdollah, T., Ganapathy, S., Holley, K.: Soma: A method for developing service-oriented solutions. IBM Systems Journal **47**(3), 377–396 (2008)
3. Bano, M.: Aligning services and requirements with user feedback. In: RE 2014, pp. 473–478 (2014)

4. Bano, M., Zowghi, D.: Users' voice and service selection: an empirical study. In: EmpiRE 2014, pp. 76–79 (2014)
5. Bano, M., Zowghi, D., Ikram, N., Niazi, M.: What makes service oriented requirements engineering challenging? A qualitative study. IET Software 8(4), 154–160 (2014)
6. Casamayor, A., Godoy, D., Campo, M.: Identification of non-functional requirements in textual specifications: A semi-supervised learning approach. IST 52, 436–445 (2010)
7. Cleland-Huang, J., Gotel, O.C.Z., Huffman Hayes, J., Mäder, P., Zisman, A.: Software traceability: trends and future directions. In: FSE 2014, pp. 55–69. ACM (2014)
8. och Dag, J.N., Gervasi, V., Brinkkemper, S., Regnell, B.: A linguistic-engineering approach to large-scale requirements management. IEEE Software 22, 32–39 (2005)
9. Ferrari, A., Lipari, G., Gnesi, S., Spagnolo, G.O.: Pragmatic ambiguity detection in natural language requirements. In: AIRE 2014, pp. 1–8, August 2014
10. Ferrari, A., Gnesi, S.: Using collective intelligence to detect pragmatic ambiguities. In: RE 2012, pp. 191–200. IEEE (2012)
11. Gervasi, V., Zowghi, D: Supporting traceability through affinity mining. In: RE 2014, pp. 143–152. IEEE (2014)
12. Huergo, R.S., Pires, P.F., Delicato, F.C., Costa, B., Cavalcante, E., Batista, T.: A systematic survey of service identification methods. SOCA 8(3), 199–219 (2014)
13. Jamshidi, P., Khoshnevis, S., Teimourzadegan, R., Nikravesh, A., Shams, F.: Toward automatic transformation of enterprise business model to service model. In: PESOS 2009, pp. 70–74. IEEE Computer Society (2009)
14. Kontogogos, A., Avgeriou, P.: An overview of software engineering approaches to service oriented architectures in various fields. In: WETICE 2009, pp. 254–259. IEEE (2009)
15. Manning, C.D., Raghavan, P., Schütze, H.: Introduction to Information Retrieval, Volume 1. Cambridge University Press (2008)
16. Papazoglou, M.P., Traverso, P., Dustdar, S., Leymann, F.: Service-oriented computing: a research roadmap. IJCIS 17(02), 223–255 (2008)
17. Salton, G., Wong, A., Yang, C.-S.: A vector space model for automatic indexing. Communications of the ACM 18(11), 613–620 (1975)
18. Settle, R.B., Golden, L.L.: Consumer perceptions: Overchoice in the market place. Advances in Consumer Research 1, 29–37 (1975)
19. Tan, P.-N., Steinbach, M., Kumar, V.: Introduction to Data Mining. Addison-Wesley (2005)
20. Turney, P.D., Pantel, P.: From frequency to meaning: Vector space models of semantics. J. Artif. Intell. Res. 37, 141–188 (2010)
21. Wirsing, M., Hölzl, M.: Rigorous Software Engineering for Service-oriented Systems: Results of the SENSORIA Project on Software Engineering for Service-oriented Computing, vol. 6582. pringer, Heidelberg (2011)
22. Zachos, K., Maiden, N.A.M.D., Howells-Morris, R.: Discovering web services to improve requirements specifications: does it help? In: Rolland, C. (ed.) REFSQ 2008. LNCS, vol. 5025, pp. 168–182. Springer, Heidelberg (2008)
23. Zadeh, A.T., Mukhtar, M., Sahran, S., Khabbazi, M.R.: A systematic input selection for service identification in smes. Journal of Applied Sciences 12(12), 1232–1244 (2012)

Model-Based Approach for Engineering Adaptive User Interface Requirements

Kibeom Park[1](✉) and Seok-Won Lee[2]

[1] Graduate School of Software, Ajou University, Suwon, Republic of Korea
pkb@ajou.ac.kr
[2] Department of Software Convergence Technology, Ajou University,
Suwon, Republic of Korea
leesw@ajou.ac.kr

Abstract. Although Model-Based User Interface (MBUI) design approaches have been suggested and researched over a long period of time, the advantages of adopting them into the development of Adaptive User Interface (AUI) have not stood out. We believe that it is due to the lack of an integration of the Requirements Engineering (RE) process, and methodologies for Model-based AUI development. Since RE provides a solid base to the development of software, requirements of AUI have to be preceded appropriately in the development process. Previously, we suggested a RE method for AUI reflecting the viewpoint of Self Adaptive System (SAS). In this paper, we elaborate on our previous method grounded on a model-based approach. The proposed method is illustrated with an example scenario, which makes adaptations of the user interface at run-time by conforming to the context of users. Finally, an evaluation of our method is provided by a case study at the end of the paper.

Keywords: Adaptive User Interface (AUI) · Model-Based User Interface (MBUI) · Requirements Engineering (RE) · Self-Adaptive System (SAS) · User Interface (UI)

1 Introduction

Due to the widespread popularity of mobile devices, demands of end users to be provided with personalized and customized services have been tremendously raised these days. For achieving such a personalization of services, a system should fulfill dynamic requirements varying in different contexts at runtime. One of these attempts to support such runtime personalization is realizing the use of Adaptive User Interface (AUI).

AUI is a user interface (UI) that has the ability to adapt itself by reasoning a suitable presentation of the service according to a situation at run-time. Realizing AUI is considered a difficult challenge because a system needs to monitor and analyze the context data of a user and his or her environment at run-time without any decisions made by human. It demands several pre-defined rules and a knowledge base to make proper adaptations of UI to satisfy diverse user demands.

© Springer-Verlag Berlin Heidelberg 2015
L. Liu and M. Aoyama (Eds.): APRES 2015, CCIS 558, pp. 18–32, 2015.
DOI: 10.1007/978-3-662-48634-4_2

Fig. 1. UI adaptation prototype when user is standing (left) and running (right)

Fig. 1 illustrates a simple prototype of AUI in healthcare mobile application. The left side, which represents a normal version of UI, has several UI components; a title, sub-title, picture, texts with a small-sized font, and two buttons. One of the requirements is that the screen displayed to the user should be adequately changed with different user activities. For example, the normal version of UI is no longer usable when the user is running, since the user finds it hard to recognize contents in the screen due to the wobbling motion produced.

In this case, the UI can be adapted by increasing the font size and hiding some of the unimportant content. One possible adaptation is provided on the right side. If we assume that the subtitle and two buttons at the bottom are important, we can not only hide picture and texts with a small-sized font (considered unimportant content), but also increase the size of the subtitle and two buttons (important content). By doing so, we now can preserve usability, readability and legibility. Fig 1. will be used again for describing our method in section 3.

It is obvious that we cannot consider AUI just as a change of 'views', since it requires considerable thoughts in 'logic' for monitoring context data, reasoning suitable versions of UI adaptation, etc. Therefore, it brings us to look at AUI from the domain of software engineering.

For making a design of AUI, Model-based User Interface (MBUI) design approaches such as CAMELEON-RT [1] and several User Interface Description Languages (UIDLs) like UIML[2] and UsiXML[3] have been proposed for decades in the Human Computer Interaction (HCI) research area. Even though these approaches enable software engineers to develop infrastructure for AUI, it is still hard to apply them to actual AUI development so far. One of the major reasons is the lack of consideration of Requirements Engineering (RE). Since RE provides a solid base to the development of software, requirements of AUI have to be preceded appropriately in the development process.

In our previous work [4], we proposed requirements elicitation guidelines using general concepts of a Self-Adaptive System (SAS). However, we did not adopt existing MBUI design research from the HCI area. There continues to be a gap between RE for AUI in the software engineering area and MBUI in the HCI area.

In this paper, we extend our work by introducing the viewpoints of model-driven development. We exploit the MBUI research, and accordingly, the benefits of MBUI approaches are adopted into the RE method for AUI.

The rest of the paper is organized as follows: In section 2, we review the related literature in three parts. The definition and the characteristics of AUI are surveyed in the first part, and the previous MBUI design approaches are described in the second part. We then review our previous requirements elicitation guidelines for AUI in the last part. In section 3, we proposed an extended RE method for AUI by adopting a model-based approach and show how our method works by illustrating an example from the mobile application domain. Section 4 provides an evaluation of our method by application to a case study. At last, we conclude the paper with the summary of contributions and future works in section 5.

2   Literature Review

2.1   Adaptive User Interface

As we mentioned above, AUI is the UI that has an ability to adapt itself to the change of the context of user and device. The term 'context' as used above can vary by many design decisions of the developers. It can be the static context of individual user or device, including user characteristics [5], the user's physical capability, user preference, device type, etc. It can be a dynamic context considering spatial or temporal factors as well.

The term 'adaptive' should be distinguished from the term 'adaptable' [6]. One of the main differences between Adaptable UI and Adaptive UI is that Adaptable UI uses only the static context of the user while Adaptive UI uses both the static and dynamic contexts of the user. Fig. 2 portrays the difference between those two approaches and their relationships with static context and dynamic context. This shows that AUI is much broader concept than Adaptable UI.

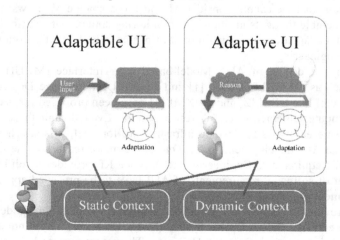

Fig. 2. Difference between Adaptable UI and Adaptive UI

Adaptable UI requires user input on such things including preference, physical capability, etc. The system then makes suitable UIs based on the user inputs. For instance, imagine a system that requires user input on whether the user is visually impaired or not, and then uses the response as a parameter for the proper adaptation. If the user has difficulty seeing, a voice-guided UI might be provided to the user in this case. Another example is that the UI that changes itself depending on the device type, such as display size. The system may choose a suitable version of UI for current display size of the device amongst already designed UIs. Even though the system could figure out the display size of the current device, it would be categorized as making a choice based on user inputs among the design templates during the design time.

AUI, on the other hand, contains the intention of generating a design at run-time rather than choosing a design at the design time. Instead of requiring user inputs, it examines the context of the user itself. Therefore, AUI reduces the burden of the user to respond in advance, unlike Adaptable UI. Furthermore, it makes reasoning about the situation and adapting the UI at run-time possible.

2.2   Model-Based User Interface (MBUI) Design

MBUI approaches have been proposed for decades as a way of achieving a technical basis for realizing AUI. The key concept is to separate UI development in multiple layers. One insightful paper addressing this concept is [7], which suggested the model-based UI framework with three layers: an Abstract UI, Concrete UI and a Final UI.

Abstract UI describes what the user actually works with the system. Currently, in the interactive system research area, researchers use a Task model and Domain model for description of Abstract UI. A Task model represents a task flow of user interactions, and a Domain model describes the knowledge of UI components which later can be used for adaptation.

Concrete UI describes more specific graphical representation based on Abstract UI. In this layer, the UI is dependent on platform or devices. It can be represented as a high-level user interface description languages (UIDL) such as UIML[2], UsiXML[3], etc. It can contain several UI design alternatives since it is written in high level.

Final UI describes final UI design decision. In this layer, high-level Concrete UI design is specified in more details. One of the examples of Final UI is a choice between 'radio button' or 'select box', in case the Concrete UI denotes 'graphical selection' as a high-level design decision.

Although an MBUI approach has been suggested and researched over a long period of time, the advantages of adopting it in the development of AUI did not stand out enough. A model-based approach in UI development focuses on solving device compatibility, but does not mention how they could be used in the development of AUI. We insist that this problem is caused by a lack of the RE process and methodologies for AUI. Currently, the requirements are addressed only in Abstract UI and the models in this layer provide much too abstract requirements. For this reason, detailed requirements that should exist according to specific domains are ignored and not addressed properly.

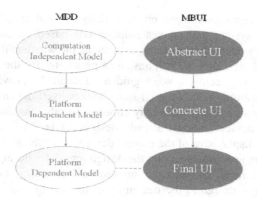

Fig. 3. A diagram mapping the relationship between Model-driven Development (MDD) and Model-based User Interface (MBUI)

In this paper, we strongly believe that the strong points of adopting model-based approach are also found in the development of AUI. A RE method, especially, can be elaborated with such a kind of approach. The following are the benefits of MBUI approach, which are normally addressed in the UI development. We will use these criteria to evaluate our method, which benefits from these in section 4.

A. Reusability

Model-based approaches in UI development enable the reuse of models, meta-models and transformations of UI components by separating the concerns of each level. When the needs are changed, developers can easily choose another way of specific adaptation alternatives easily without changing the high-level design. In this case, developers are able to reuse designs available in high-level layers.

B. Run-time Adaptivity

Making different adaptations at run-time is most suitable since the layered model explicitly structures a higher level goal and its related lower level design alternatives accordingly. This makes a system knowledgeable to what to adapt in a specific way to satisfy the high-level goal.

C. Modifiability

One of the benefits of a MBUI approach is that it enables a choice between an alternative design or to change the current design. The reason is that the designs are separately made with three layers according to the level of abstraction, and then it separates the concerns in the design. With layers, the detailed adaption logic can be modified easily.

2.3   Requirements Elicitation Method for Adaptive User Interface

Requirements elicitation is the first step in drawing the needs of users and various stakeholders. Since AUI requires motives triggering UI adaptation and several logics for monitoring the context at runtime, it is obvious that the requirements for AUI should be clearly elicited and well defined.

In fact, UI development has not been considered in the RE area well, as it has been regarded more in the domain of HCI area. Also, researchers of MBUI have developed different methodologies and terms that make it difficult to integrate that research into a software engineering perspective. One example is that researchers of developing Interactive System use different methodologies like a 'Task model' and 'Domain model' instead of RE for handling the goal and flow of UIs. Although there exist some research using those methodologies as a basis for RE [8], they do not expand its coverage towards AUI. There continues to be a lack of research addressing adequate RE methodologies for AUI.

For this reason, our previous work focused on how to elicit initial requirements of AUI [4]. We proposed guidelines for eliciting AUI requirements using well-known concepts from SAS research.

Our previous paper introduced a 3-step requirements elicitation process including 'AREA – BASE – CONSEQUENCE', focusing on AUI development in the domain of the mobile application. In the first step 'AREA', we considered context, property and constraints of the domain. Then we re-defined the term 'MAPE-K Loop' [9], 'Self-* Properties' [10], and their elements considering the contents of AREA for making them proper to the domain of mobile application. During the second step

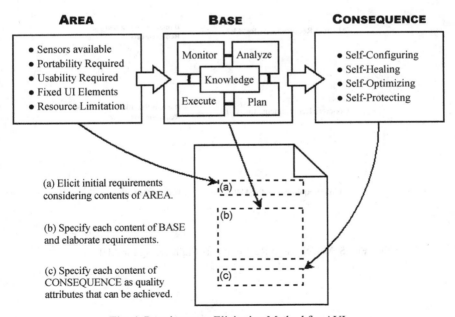

Fig. 4. Requirements Elicitation Method for AUI

'BASE', developers should use each element of MAPE-K Loop 'Monitor', 'Analyze', 'Plan', 'Execute' and 'Knowledge' as criteria for eliciting functional requirements. In this way, requirements could be elaborated more upon according to different domains. During the last step, 'CONSEQUENCE', re-interpreted Self-* Properties are used as a guideline for eliciting quality attributes. These elements could be the minimum quality attributes for achieving an adaptivity. That is, Self-Configuring, Self-Healing, Self-Optimizing, Self-Protecting should at least be elicited.

Previously, our work did not address how to bridge elicited requirements to MBUI design approaches. If we integrated the merits of MBUI, we could exploit many benefits of an MBUI approach in the development of AUI. In this reason, we extend our previous method by introducing a model-based approach in the paper.

3   Model-Based Engineering for AUI Requirements

3.1   Adopting Model-Based Approach to AUI Requirements

Fig. 5 shows how we can adopt the notion of model-driven development into RE. Three layered modeling from [7] is adopted to RE, respectively: Abstract Requirements for AUI, Concrete Requirements for AUI, and Final Requirements for AUI.

Abstract Requirements for AUI represent domain-independent AUI requirements while Concrete Requirements for AUI represent domain-specific requirements of AUI. Final Requirements for AUI mean detailed variation on UI elements.

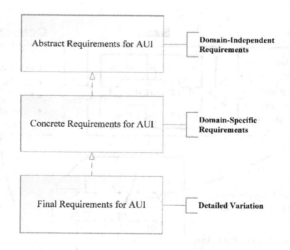

Fig. 5. Model-based Requirements Engineering for AUI

3.2   Proposing an Extended Method for AUI

In this section, we propose an extended requirements elicitation method for AUI by bringing a model-based approach into RE. Our previous work is restructured with a three-layered model-based notion in this paper. Fig. 6 gives an explanation of our approach. Our method guides software engineers to specify AUI requirements with three layers according to the level of abstraction.

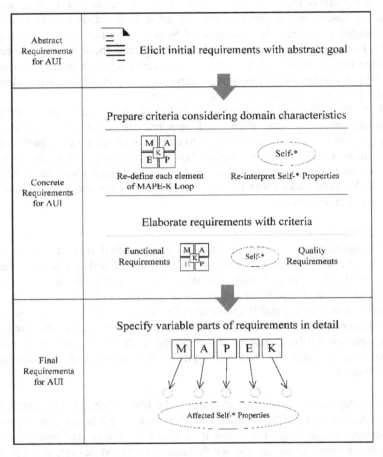

Fig. 6. Extended Requirements Engineering Method for AUI

A.   Abstract Requirements for AUI

Abstract Requirements for AUI are described with the highest upper-level goals. These requirements might correspond to the initial rough requirements with goals and intentions. The requirements at this point are general, and not tailored to specific machine or domain. Therefore, we can say that they are kept independent on the specific domain, and it can be reused when the AUI should be adopted into other domains or machines with similar purposes.

B.   Concrete Requirements for AUI

Concrete Requirements for AUI are dependent on the certain areas where software belong. In this layer, software engineers should consider the characteristics, unique features, and standards of the specific domain area. When the software engineer finds out the application domain area of software, the next thing to do is to define each element of the MAPE-K Loop and interpreting each property of Self* Properties, according to the domain. The elements of MAPE-K Loop are Monitor/Analyze/Plan/Execute, and these are used as the criteria to verify whether functional requirements are elicited well or not, in the perspective of a self-adaptive system. The elements of Self-* Properties are Self-Configuring, Self-Healing, Self-Optimizing, Self-Protecting, and these are consequences of elicited requirements which can be used to verify whether quality attributes are elicited in the perspective of self-adaptive system well or not. Each element of Self-* Properties also should be specified in the step.

Following the criteria based on MAPE-K Loop and Self-* Properties, software engineers can elicit AUI requirements easily. In addition, these requirements would be elaborated on to comply with the criteria. In this step, requirements should be more concrete than the previous step, but must not be too specific as to disturb the modification of adaptation rules. Concrete requirements for AUI should make room for the changeable part for the adaptation so that it preserves modifiability.

C.   Final Requirements for AUI

We mentioned that the requirements that are specified in the previous steps do not contain too specific of requirements, such as detailed adaptation rules or UI elements to be adapted, in order to make a room for changeable parts. In this final layer, detailed requirements that can be expected to be changed later are specified and analyzed. Final Requirements for AUI are domain/machine-dependent requirements, and these can also cause changes of the quality attributes. Therefore, quality attributes specified complying with Self-* Properties might be changed accordingly. Related quality attributes are connected and traced through higher layers.

The three steps above, which are divided by the level of abstraction, enable the separation of the concerns in the requirements. Not only does it take the strong points of a model-based approach, it also makes it easy to trace the related requirements and change parts without changing the whole requirements. Moreover, this can be extended to each design level. We will later show the strengths of our method.

3.3   Method Illustration with an Example Scenario

For a deeper understanding of our method, we now illustrate our method with an example scenario. We elicit, analyze, and specify AUI requirements by following our model-based RE method for AUI. The results correspond to the requirements for implementation of the right side of Fig. 1.

The AUI scenario in our example is in the situation of the development of a health-care mobile application. We assume that users might be everywhere, including both inside and outside of their home. Also, users might have different characteristics and ages, genders, characters, jobs, etc. The application can satisfy the personalized demands of users by adopting AUI in the application, considering both the user's situation and environments.

We assume that the initial goal to achieve from AUI is to make adaptations of UI at runtime to conform to the context of users. For achieving our initial goal, initial requirements from various stakeholders should be elicited. As one example, we would use the following requirement in the Abstract Requirements for AUI layer: *If the user finds it hard to see UI clearly because of the changes of environment, AUI increases its usability by changing its UI.*

In the Concrete Requirements for AUI layer, in order to make unclear requirements clear, the characteristics and constraints in the domain area of the software are considered. For example, Table 1 represents such considerations in the development of the mobile application. In our previous paper, we elicited considerations of the mobile domain from [11].

Table 1. Considerations of Mobile Domain

Consideration	Description
Sensors Available	Several sensors including an accelerometer are available.
Portability Demand	It should be compatible among multiple platforms/machines.
Usability Demand	Personalized user-centric service should be provided.
Fixed UI Elements	It should use a UI library that already exists.
Limited Resource	Resources such as battery, CPU, storage are limited.

Table 2. Re-defined Descriptions of MAPE-K Loop Elements

MAPE-K Loop	Description
Monitor	Data and its monitoring method for recognizing situation.
Analyze	Rule for analyzing situation.
Plan	Rule for adaptation.
Execute	Actual Adaptation Behavior.
Knowledge	Knowledge required for recognizing situation and adaptation.

Table 3. Re-interpreted Descriptions of Self-* Properties Elements

Consideration	Description
Self-Configuring	Personalization according to user characteristics and situation.
Self-Healing	The ability to recover the usability when unexpected increase of UI complexity occurs,
Self-Optimizing	UI optimization according to machine profile and resource(Battery, CPU, etc.) situation.
Self-Protecting	Preventing UI crashes or defending when UI crashes.

MAPE-K Loop and Self-* Properties are re-defined or re-interpreted according to the domain. In our case, we analyze them in the mobile domain. Table 2 represents the re-defined descriptions of MAPE-K Loop elements considering the mobile domain. Table 3 represents the re-interpreted descriptions of elements in the Self-* Properties.

The next step for software engineers is to specify concrete requirements according to the criteria that we have constructed. Table 4 contains the requirements specification of this step. Detailed types of sensors or rules for adaptation are not specified in this step.   That is because those parts have high possibility to change.

Detailed rules that have high chances of change are specified in the next step, the Final Requirements for AUI. In this layer, related quality attributes are revisited and modified. The specification of Final Requirements for AUI is represented in Table 4. These contain adaptation rules and situation recognition rules which can be modified often due to the demands of stakeholders.

Table 4. Abstract, Concrete and Final Requirements of AUI

Abstract Req.	If the user finds it hard to see UI clearly because of the changes of environment, AUI increases usability by changing its UI.			
Concrete Req.	**Monitor**	**Analyze**	**Plan**	**Self-* Property**
	The AUI monitors wobbles (shakes) through sensors that Android phone provides.	The AUI analyzes data and determines whether the user is interrupted by current wobble or not.	The adaptation rule is a UI simplicity rule that is performed when AUI determines that usability is getting too lower.	Self-Configuring, Self-Healing, Self-Optimizing
	Execute		**Knowledge**	
	Plan is actually performed when the decision is triggered.		Sensor monitoring interval, usability metrics, UI simplicity rule, Adaptation rule.	
Final Req.	**Monitor**	**Analyze**	**Plan**	
	The type of sensors that is used: Gyroscope	Usability evaluation algorithm: When gyroscope data is above a certain level, usability value decreases.	UI simplicity rule: Increase text size and hide unimportant UI elements.	
	Execute		**Knowledge**	
	Trigger rule: When usability value is under a certain level, adaptation triggers.		Sensor monitoring interval, usability metrics, UI simplicity rule, Adaptation rule.	

4 Evaluation

In this section, we design a case study for evaluating our method by following the case study design methodology in [12].

4.1 Study Questions

Our model-based RE method for AUI in this paper also uses the advantages of using model-based approach in the UI development that we described in the section 2.2. We previously described three benefits of using a model-based approach: Reusability, run-time adaptability and modifiability. Our proposed method also introduces the traceability. In addition, this makes requirements to be easily extended to the design. To prove that these advantages are obtained by using our method, following questions are discussed.

- *Q1. Does it support reusability?*
- *Q2. Does it support run-time adaptivity?*
- *Q3. Does it support modifiability?*
- *Q4. Does it support traceability?*
- *Q5. Is it extendable to design level?*

4.2 Case Study

Table 6 shows the evaluation of our method by answering each question. Each answer shows whether it is enough to support each question, and if so, the evidence that supports it.

Table 5. Method Evaluation

Method Evaluation Table			
	Study Question	*Support*	*Evidence*
Q1	Does it support reusability?	Yes	When the domain or machine targeting is changed, requirements of upper layer can be used.
Q2	Does it support run-time adaptivity?	Yes	It controls UI elements separately since it can have knowledge about adaptation.
Q3	Does it support modifiability?	Yes	When adaptation rules should be changed, software engineers can change only local parts, instead of changing the whole part, because it is only related to the Final requirements.
Q4	Does it support traceability?	Yes	Requirements among layers and their affected quality attributes are easily found.
Q5	Is it extendable to design level?	Yes	Design and implementation are performed from requirements.

A. Reusability

It is reusable because it separates the concerns into three layers, and upper layers are still preserved when changes need to occur locally. For example in our scenario, when software engineers decide to support a desktop application, they can reuse Abstract Requirements for the desktop application.

B. Run-time Adaptivity

It supports run-time adaptivity because it allows AUI to have knowledge about the adaptation. Unlike traditional adaptable UI, which only allows for a few versions of whole UI, our method supports the development of a lot of versions of AUI. That is because our method can control each UI element individually and it is possible to adopt more detailed adaptations. For example, UI simplicity rules contain knowledge about the importance of each UI element, which enables separate control of each element.

C. Modifiability

It is modifiable because it separates the concerns into three layers, and without changing the whole AUI, software engineers can change only the local part in question. For example in our scenario, when software engineers decide to change the UI simplicity rule, they can specify only the Final Requirements for AUI again.

D. Traceability

Traceability is ensured because it has been derived from requirements from the upper layer. For example, when Final requirements should be changed, corresponding Concrete Requirements of AUI and their quality attributes are easily found.

E. Extendibility to Design

The model-based requirements can be easily extended to design and implementation. We designed the logic of the simple prototype of Android application, and it is shown in Fig. 1. In our design, Concrete Requirements for AUI are made to a Java Interface that includes Monitor, Analyze, Plan, Execute as the methods, and Knowledge as variables. Then, we implement FinalUI, which satisfies the ConcreteUI.

```
package pkb.nise.ajou.ac.kr.adaptiveuserinterface.aui;

/**
 * Created by AJOU on 2015-06-25.
 */
public interface ConcreteUI {
    public void startMonitoringSensor();
    public void stopMonitoringSensor();
    public void analyzeUserMoveStatus();
    public void planToSimplifyUI();
    public void executeAdaptation();
}
```

Fig. 7. Extending Concrete Requirements to the Design of Java Interface

Through the design and implementation, we found that our method allows for extending requirements easily to design and implementation. Fig. 7 shows the possibility of extending the design.

5 Conclusion and Discussions

Requirements elicitation process and methodologies for AUI development have not been defined well so far. Although our previous work suggested a guideline for AUI requirements, it did not reflect the advantages of MBUI design approaches.

In this paper, we proposed an extended RE method for AUI by adopting a model-driven approach. We showed how our method works by illustrating case study examples. By using our method, software engineers can effectively elicit requirements in AUI development step by step.

However, this paper contains several points of discussion, which can be seen as follows. First, the approach we suggest is still at a very high level or can be considered too general, in that many details are omitted. There remain questions about how to transform requirements in different layers. Furthermore, RE processes after the elicitation is not addressed well. In the future, we plan to focus on this question to elaborate our method.

Second, the evaluation is not enough since we conducted only one case study. In addition, the metrics for evaluation hold some threats to its validity. We will conduct more case studies for a better evaluation in the future.

Lastly, it is not clear how existing RE approaches can be integrated into the proposed framework or design approaches. For example, there is already much research that address the model-based AUI design approach such as that in [13]. We will bridge our RE method to the design method of AUI by further research.

Acknowledgment. This research was supported by the Next-Generation Information Computing Development Program through the National Research Foundation of Korea (NRF) funded by the Ministry of Science, ICT & Future Planning (2013M3C4A7056233).

References

1. Balme, L., Demeure, A., Barralon, N., Calvary, G.: CAMELEON-RT: a software architecture reference model for distributed, migratable, and plastic user interfaces. In: Markopoulos, P., Eggen, B., Aarts, E., Crowley, J.L. (eds.) EUSAI 2004. LNCS, vol. 3295, pp. 291–302. Springer, Heidelberg (2004)
2. Abrams, M., Phanouriou, C., Batongbacal, A.L., Williams, S.M., Shuster, J.E.: UIML: an appliance-independent XML user interface language. Computer Networks **31**(11), 1695–1708 (1999)
3. Limbourg, Q., Vanderdonckt, J., Michotte, B., Bouillon, L., López-Jaquero, V.: USIXML: a language supporting multi-path development of user interfaces. In: Feige, U., Roth, J. (eds.) DSV-IS 2004 and EHCI 2004. LNCS, vol. 3425, pp. 200–220. Springer, Heidelberg (2005)

4. Park, K., Lee, S.W.: Requirements Elicitation for Mobile Adaptive User Interface based on Concepts from Self-Adaptive Software. In: Korea Conference on Software Engineering 2015. Korean Institute of Information Scientists and Engineers Software Engineering Society (2015)
5. Kühme, T.: A user-centered approach to adaptive interfaces. In: Proceedings of the 1st International Conference on Intelligent user Interfaces, pp. 243–245. ACM (1993)
6. Stephanidis, C., Paramythis, A., Sfyrakis, M., Stergiou, A., Maou, N., Leventis, A., Karagiannidis, C.: Adaptable and adaptive user interfaces for disabled users in the AVANTI project. In: Trigila, S., Mullery, A., Campolargo, M., Vanderstraeten, H., Mampaey, M. (eds.) Intelligence in Services and Networks: Technology for Ubiquitous Telecom Services., vol. LNCS, pp. 153–166. Springer, Heidelberg (1998)
7. Calvary, G., Coutaz, J., Thevenin, D., Limbourg, Q., Bouillon, L., Vanderdonckt, J.: A unifying reference framework for multi-target user interfaces. Interacting with Computers 15(3), 289–308 (2003)
8. Reichart, D., Forbrig, P., Dittmar, A.: Task models as basis for requirements engineering and software execution. In: Proceedings of the 3rd annual conference on Task models and diagrams (pp. 51-58). ACM (2004)
9. Kephart, J.O., Chess, D.M.: The vision of autonomic computing. Computer 36(1), 41–50 (2003)
10. Salehie, M., Tahvildari, L.: Self-adaptive software: Landscape and research challenges. ACM Transactions on Autonomous and Adaptive Systems (TAAS) 4(2), 14 (2009)
11. Wasserman, A. I.: Software engineering issues for mobile application development. In: Proceedings of the FSE/SDP Workshop On Future Of Software Engineering Research, pp. 397-400. ACM (2010)
12. Lee, S. W., Rine, D. C.: Case Study Methodology Designed Research in Software Engineering Methodology Validation. In: SEKE, pp. 117-122 (2004)
13. Akiki, P. A., Bandara, A. K., Yu, Y.: Adaptive model-driven user interface development systems. ACM Computing Surveys, 47(1), In-press (2015)

A Framework for Data-Driven Automata Design

Yuanrui Zhang, Yixiang Chen[(✉)], and Yujing Ma

MoE Engineering Research Center for Software/Hardware Co-design Technology
and Application, Software Engineering Institute, East China Normal University,
Shanghai 20062, China
yxchen@sei.ecnu.edu.cn

Abstract. The traditional model-driven developing methods in requirement engineering (RE) have met challenges. Under the dropback of big
data, we propose a new framework of software design method based
on requirement data. Given a set of requirement data, which are usually acquired directly from users' intuitive descriptions about systems,
we consider the method to analyze them and extract useful information
from them in order to build the formal specifications of systems. Here
the 'data' could be any form and could describe any prospect of functionalities of systems, the 'model' could be any types of formal models like
process algebra, automata, or some forms of state-diagrams. We limit the
scope of our discussion in this paper to a special type of data and formal
models. We first use some simple examples to clarify the general idea of
'data-driven' we propose. Then as a case study, we apply this idea to the
requirement specification of a special type of systems—spatial-temporal
systems by proposing a special formal model in order to capture the
spatial-time features of them.

Keywords: Requirement engineering · Data-driven · Big data ·
Automata theory · Spatial-temporal · Spatial hybrid automata

1 Introduction

In software engineering, as system is becoming more and more complex in size
and logic, traditional methods for software developing like event-driven design
or object-oriented design are becoming more and more difficult to meet our
requirements. We need more convenient and intuitive way to design our software
while the software we design should fully satisfy our requirements. Recently, big
data has become a hot topic in computer science. As data sets grow rapidly in
size, people start thinking about new ways of analyzing, capturing and extracting
useful information from data sets. This big change lead us to think of a new way
to develop our software as big data can also offer large information about the
requirements of systems.

Many developing methods of software engineering have been investigated
throughout the history. Niklaus Wirth gives the well-known formula: Program =
Data structure + Algorithm. This formula has guided us many years to develop

© Springer-Verlag Berlin Heidelberg 2015
L. Liu and M. Aoyama (Eds.): APRES 2015, CCIS 558, pp. 33–47, 2015.
DOI: 10.1007/978-3-662-48634-4_3

software. Later, one develops the formalization method to establish the (behavior) semantics and relations of (between) programs or data structure. Finite state machine is a classical method to describe such semantics and relations. As the development of large-scale and complex software, one explores the model-driven design framework. UML [1] is a typical platform of model-driven software development. Finite state machine is still the core of UML for description of translation among states.

Wikipedia [2] defines that data-driven programming is a programming paradigm in which the program statements describe the data to be matched and the processing required rather than defining a sequence of steps to be taken.

Wirfs-Brock and Wilkerson in [3] give the concept of data-driven design. They think that data-driven design is the result of adapting abstract data type design methods to object-oriented programming. The data-driven approach to object-oriented design focuses on the structure of the data in a system. This results in the incorporation of structural information in the definitions of classes.

Related to data-driven design, event-driven design is proposed and originated in the area of active databases in the 1980s with the introduction of triggers to database system [4][5]. A trigger is an Event-Condition-Action (ECA) rule which is checked by the DBMS whenever an event occurs that matches the Event part of the rule [6]. In event-based systems, an event can take many forms and data are abstracted as the event instance [7]. Miro Samek in [8] says that state machines are perhaps the most effective method for developing robust event-driven code for embedded systems.

Requirements engineering (RE) [9] refers to the process of defining, documenting and maintaining requirements and to the subfields of systems engineering and software engineering concerned with this process[10][11][12][13].

Formal method is an important method for requirement specification. We proposed a process algebra called STeC for specifying of real-time system with spatial temporal consistency requirements [14].

In this paper, we propose a new developing method for software engineering based on data sets. The core idea is that we can build the specifications of systems based on the data sets which offer requirement information about systems using some mechanism. To make things easier, we restrict our forms of data to be the sequence of actions that the system behave, our models to be the abstract automata. We give some easy-understanding examples to explain our ideas. The paper is organized as follows: in Section 2, we introduce our data-driven framework and basic concepts of each components of the framework. In Section 3, we give some simple examples of data domain and their relevant formal models using classical automata theories. In Section 4 we give a case study and explain how to build formal specifications for a spatial-temporal system based on spatial-temporal requirement data using our data-driven method. We introduce a type of state machine called spatial-hybrid automata as its formal model. At last, we make a conclusion in Section 5.

2   Framework for Data-Driven Automata Design

Fig. 1 gives the core idea of data-driven framework. The data offers the requirements of systems. It can give information about how a system behave, or for instance, some features about a system. The model gives the formal specifications of systems, it needs to be well chosen and generated based on the data. The traces (generated by the model) is analyzed if it fits our requirements. If not, more data is needed to give more concrete requirements. The key part of this procedure is the mechanism to choose and generate the formal model from data, which could be hard and varies from different data domains and requirements. In our paper, we only consider the data as some types of sequences of events, and models as some kinds of automata.

2.1   Data Type, Trace and Word

A data is a set of traces that describe a sequential of events of a system. Since the types of behaviours of systems vary from one to another, the type of data should also be different. For some systems we only consider the logical relationships between their actions, for some others, we also need to consider the time issues together with actions, and for the rest, we even consider their space relationships.

We next give a definition of a data, trace and word. A trace consists of a sequence of words. A word is a tuple of different elements that belong to different domains, which makes data distinguished from one to another.

Definition 21. *A word is a n-tuple* $w \in D_1 \times D_2 \times ... \times D_n$, D_i *(i* $\in \{1, 2, ..., n\}$*) is called data domain.*

Definition 22. *A trace is a finite or infinite sequence of words, denoted as* $Tr = w_1 w_2 ... w_n ...$ *where* $w_i \in D_1 \times D_2 \times ... \times D_n$ *for all* $i \in \{1, 2, ...\}$ *and some* $n \in \mathbb{N}^+$.

Definition 23. *A data is a set of traces for some type of word, denoted as* $Data = \{Tr \mid Tr = w_1 w_2 ... w_n ..., \ w_i \in D_1 \times ... \times D_n \ \forall i \in \mathbb{N}^+\}$.

The data domain D can be any mathematical structure, e.g, $\mathbb{R}$, $\mathbb{R}^+$, $\mathbb{N}$, any partial order sets, etc.

2.2   Languages and Automata

The language in computer science is undertood as a set of words, like data, in our example. So in the data-driven framework of automata design, we can equivalently treat data as languages, with different types of word (Def. 21). For some types of languages, for example, regular languages [16], the mechanism of generating the corresponding automata is determined. That is, for any regular languages, its generated model can fully fit it. For example, determined finite automata (DFA) can be generated from regular languages [17], timed automata [18] can be generated from timed regular expressions [19], hybrid

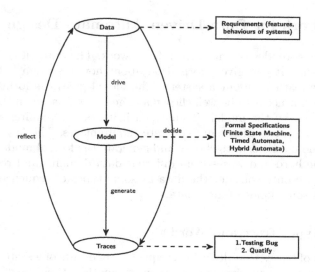

Fig. 1. The Data-Driven Framework of Software Development

automata [20] can be generated from hybrid traces, etc. The same for the context-free languages or ω-languages. However, not all languages can be easily translated into automata, for example, timed non-regular languages can not be translated into a timed automata. Also, for the languages that contains an informal type of words (not the word that can be accepted by automata), the algorithms for translations is not direct. In Section 4, we build a type of automata called spatial-hybrid automata for the model of spatial-temporal data types based on hybrid automata.

3 Some Examples of Data Domain

We give some examples [1] of different types of data and their generated formal specifications. We firstly introduce a simple real-time system, called 'Railroad Crossing System' (abbreviated as RCS), as an example. Then pick up some different types of data as examples based on different behaviours of this system.

3.1 Railroad Crossing System

Railroad Crossing System (or RCS) (shown in Fig. 2) is a dynamic scheduling system at a crossing. There are two agents: a train and a gate. The gate is at the crossing road where there is a railroad in east-west direction and a road in

---

[1] Some early examples of data-driven languages are the text-processing languages sed and AWK [15], where the data is a sequence of lines in an input stream ?C these are thus also known as line-oriented languages ?C and pattern matching is primarily done via regular expressions or line numbers.

south-north direction. The train communicates with the gate dynamically when attending to pass the crossing. The gate keeps opened when no train coming, and has to close itself if there is a train coming. We summarize the system scenario and list the behaviours of the train and the gate respectively as below:

RCS System Scenario:

- Train passes through a crossroad which installs a gate for safety.
- Normally, the gate is open so that cars and people may pass the railroad. But, if a train passes through this crossroad then the gate must be closed.
- This control between gate and train will be completed through sending and getting messages each other.

Train Behaviour:

- Train sends a message **App** to gate at the location **Lapp** and then goes to gate;
- At location **Lpass**, if the train gets the message **Cross** from gate then it passes through the gate and sends message **leaving** to gate at the location **Lleav**.
- Otherwise, the train must stop at the location **Lstop** and waits for the message **Cross** there;

Gate Behaviour:

- The gate is open for cars and people to pass through railroad.
- If the gate gets the message **App** then closes;
- If the gate is closed then sends a message **Cross** to train;
- When the gate gets a message **leaving** and does not get the message **App** then it opens.

Next we extract part of their behaviours and represent them as different types of data.

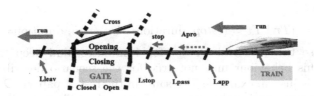

Fig. 2. Railroad Crossing System

3.2   The Data

Untimed Data. The untimed data only captures logical relationships between actions.

For example, suppose we capture the action 'open' of the gate of RCS system as a sequences, using the data defined as follows:

$$Data_{open} = \{Tr \mid Tr = \langle open \rangle^*\}$$

Tr is a trace. It is a regular languages, so the mechanism from it to its formal model is determined. Fig. 3 shows its corresponding automata.

Similarly, we can capture and generate the behaviours of sequences of 'open-close' of the gate, like:

$$Data_{OC} = \{Tr \mid Tr = (\langle open \rangle \langle close \rangle)^*\}$$

Fig. 4 shows its corresponding automata.

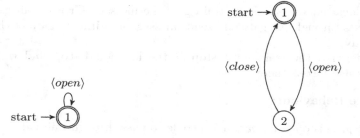

Fig. 3. Automata accepting the trace 'door open'

Fig. 4. Automata accepting the trace 'door open-close'

Timed Data. Timed data contain information about time concerning the point at which actions happen or how long actions would take through. For example, we now capture the sequences of actions of train 'stop-run' with a two-tuple arrays of data:

$$Data_{SRT} = \{Tr \mid Tr = (\langle stop, \text{duration}= 1m \rangle \langle run, \text{duration}= 5m \rangle)^*\}$$

We suppose that the train 'stop' for 1 minutes, and keep 'running' for 5 minutes...(this behaviour may not true for the scenario described above in Section 3.1, it is just an example! The same in the rest of paper). $Data_{SRT}$ contains a domain representing the duration of one action. Though it is not a formal timed language for timed automata, we can translate it into a timed language:

$$L = \{trace \mid trace = \langle stop', 1 \rangle \langle run', 6 \rangle \langle stop', 7 \rangle \langle run', 13 \rangle ...\}$$

Where we take $stop'$ as an instant action, meaning the finishing of the action $stop$, the same as run'. We can proof that such timed language L is a timed regular expression so we generate a timed automata shown as Fig. 5, we translate the 'duration' to a clock 'x' in timed automata.

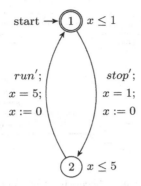

Fig. 5. Timed Automata accepting the trace 'train stop-run'

Spatio-Time Data. Spatio-time data not only have time issues, but also space issues. They record behaviours that concern both time and locations.

Suppose that the train in RCS system approach the gate at time 14:34, location 'Lapp', taking 2 minutes, then arrive at location 'Lpass' at time 14:36. It then take 8 minutes to stop at time 14:44. After waiting for 2 minutes, the train passes the gate at time 14:46, taking 2 minutes. We give the corresponding data as follows:

'Train Passing' spatio-time data:

$$Data_{TP} = \{Tr\},$$

where

$$Tr = \langle\ appr,\ \text{location}=\ Lapp,\ \text{time}=\ 14:34,\ \text{duration}=\ 2m\rangle\ \langle\ stop,$$
$$\text{location}=\ Lpass,\ \text{time}=\ 14:36,\ \text{duration}=\ 8m\rangle\ \langle\ wait,\ \text{location}=\ Lstop,$$
$$\text{time}=\ 14:44,\ \text{duration}=\ 2m\rangle\ \langle\ pass,\ \text{location}=\ Lstop,\ \text{time}=\ 14:46,$$
$$\text{duration}=\ 2m\rangle$$

See that $Data_{TP}$ is not in a form that can be accepted by any kinds of automata. So we need to make a translation, we translate data $Data_{TS}$ into a timed trace and a propositional trace of timed automata as follows:

$$trace = \langle\ appr,\ 14:34\ \rangle\ \langle\ stop,\ 14:36\rangle\ \langle\ wait,\ 14:44\rangle\ \langle\ pass,\ 14:46\rangle\ \langle\ \emptyset,$$
$$14:48\rangle$$
$$proptrace = \langle\emptyset\rangle\ \langle Lapp\rangle\ \langle\ Lpass\ \rangle\ \langle\ Lstop\ \rangle\ \langle\ Lstop\ \rangle\ \langle\ \emptyset\ \rangle$$

we now can generate timed automata according to them by adding propositions on each state representing the locations. Fig. 6 is the corresponding timed automata, the data 'location' is translated to atomic propositions in timed automata.

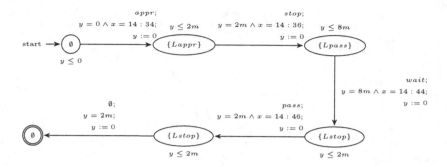

Fig. 6. Timed Automata accepting the trace 'Train Passing'

Hybrid Data. Hybrid data describes the behaviours with time issues and the change rate of variables.

Suppose that after the gate receive the message 'App' from the train, it will process for 2 seconds, then it will close itself for 30 seconds, with the rate of 3 degrees per second. After getting the message 'Leave' from the train for 2 seconds, it will open itself again for 30 seconds, with the same rate. We give the data as follows:

'Gate Close-Open' hybrid data:

$Data_{GCO} = \{Tr \mid Tr = (\langle App?, \text{duration}= 2s\rangle \langle close, \text{angle0}= 90d, \text{angle1}= 0,$
$\text{rate}= -1\rangle \langle open, \text{angle0}= 0d, \text{angle1}= 90d, \text{rate}= 1\rangle \langle Leave?, \text{duration}= 2s\rangle)^*\}$

We can translate the data into a hybrid trace and a differential equation trace as follows:

$$trace = \langle App'?, a = 90, t = 0\rangle\langle close', a = 90, t = 2\rangle\langle\emptyset, a = 90, t =$$
$$30\rangle\langle Leave'?, a = 0, t = 30\rangle\langle open', a = 0, t = 2\rangle\langle\emptyset a = 90, t = 30\rangle...$$
$$vftrace = \langle\dot{a} = 0, \dot{t} = 1\rangle\langle\dot{a} = 0, \dot{t} = 1\rangle\langle\dot{a} = -1, \dot{t} = 1\rangle\langle\dot{a} = 0, \dot{t} = 1\rangle\langle\dot{a} = 0, \dot{t} =$$
$$1\rangle\langle\dot{a} = 1, \dot{t} = 1\rangle\langle\dot{a} = 0, \dot{t} = 1\rangle...$$

Variable t records time eclapse and d records the angle value. Fig. 7 shows the generated automata.

4 Case Study—Data-Driven Method for Spatial-Temporal System

In this section, we give an example of building a formal specification for a special type of system—spatial temporal system based on a type of spatial-time data. We propose a formal model called 'spatial-hybrid automata for the specification.

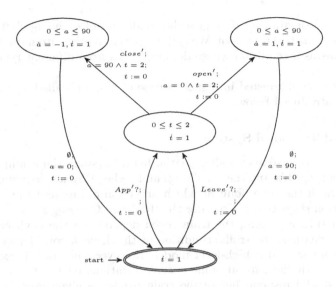

Fig. 7. The Hybrid Automata of 'Gate Close-Open'

4.1 A Counter Example

From Section 3, we show some examples of data which can generate an automata. Now consider an example of data which can not (or say not suitable) generate an automata. We change $Data_{TP}$ into a new form as:

$$Data'_{TP} = \{Tr\}$$

where

$Tr = \langle\ appr,\ \text{l}= 1km,\ \text{time}= 14 : 34,\ \text{duration}= 2m\rangle\ \langle\ stop,\ \text{l}= 0.8km,$ time= $14 : 36,$ duration= $8m\rangle\ \langle\ wait,\ \text{l}= 0.3km,\ \text{time}= 14 : 44,\ \text{duration}= 2m\rangle$ $\langle\ pass,\ \text{l}= 0.3km,\ \text{time}= 14 : 46,\ \text{duration}= 2m\rangle$

We replace the locations of each word in $Data'_{TS}$ with an exact number l. So the data turns into a hybrid data we can translate it into a hybrid trace as follows:

$trace = \langle\ appr,\ T = 14 : 34, l = 1, d = 2\ \rangle\ \langle\ stop,\ T = 14 : 36, l = 0.8, d = 8\rangle\ \langle$ $wait,\ T = 14 : 44, l = 0.3, d = 2\rangle\ \langle\ pass,\ T = 14 : 46, l = 0.3, d = 2\rangle\ \langle\ \emptyset,$ $T = 14 : 48, l = 0.3, d = 2\rangle$

$trace = \langle stop, T = 14 : 34, l = 10, d = 2\rangle\langle run, T = 14 : 36, l = 10, d =$ $\delta_1\rangle\langle stop, T = 15 : 10, l = 25, d = 2\rangle\langle run, T = 15 : 12, d = \delta_2\rangle\langle stop, T = 16 :$ $15, l = 50, d = 2\rangle...$

Now the problem rises, since the data does not specify the change rate of l, we can not generate a hybrid automata for it. It tells us we need to introduce an 'abstract' automata that only give the conditions of each edge but leave the different equations free unknown. We call it 'spatial hybrid automata' because it leave locations as a condition whose domain varies from different types of data we choose for it.

Indeed, we are interested in a special type of systems called 'spatial hybrid systems, as introduced below.

4.2   Spatial Temporal System

A type of software systems is called spatial hybrid system, the examples of this kind of systems are like Internet of Things and Cyber-Physical Systems. In their behaviours, both the time point at which an action occurs and the location at which an action stays count. Consider the 'Railroad Crossing System' (RCS) in Section 3, in such a system, the train speed depends on the environment (for example, the weather, the traffic condition or the driver), considerations which are partially ignored in a high-level model. However, the train is expected to communicate with the gate at some specific locations on the track. So is the gate, which should respond before the train reaches a given point, if to avoid any accident and have an 'optimal' use. For this type of systems, we propose a hybrid automata in a special domain—'space' domain based on hybrid automata to capture their behaviours.

4.3   Spatial Hybrid Automata

As an abstraction of hybrid automata, spatial hybrid automata add a set of special conditional triggers called 'locations', and reduce the expressive power to be weaker than hybrid automata. Such automata are good for abstracting the spatial-temporal features of spatial-temporal systems, and on the other hand though it loses the power of hybrid automata, but can be easily refined to be concrete hybrid automata.

In spatial hybrid automata, we stress that the trigger of each edge is depending on not only the conditions of variables, but also the conditions in the special triggers we define—'locations'. We take differential equations in each state as parameters and thus they are not certain. In other words, we only give the conditions ('location') when each edge can be triggered and do not tell the exact form of differential equations for each state to reach those conditions. We remain the time variables and continuous variables the same in hybrid automata, but adding our set—'locations' as part of guard conditions. The formal definition is as follows:

Definition 41. *A Spatial hybrid automaton sHa is a collection $(Q, X, L, M, \Sigma, F, Init, D, E, G, R)$, $(q, x) \in Q \times X$ as the state of sHa denoted as St, where*

1. *$Q = \{q_1, q_2, \ldots\}$ is a set of discrete states;*
2. *$X = \mathbf{R}^n$ is a set of continuous states;*
3. *$L = \{l_1, l_2, \ldots, l_n\}$ is a finite set of locations, $l_i \in P(X)$ for any $1 \leq i \leq n$ ($P(X)$ denotes the power set of X).*

4. $M= \{m_1, m_2, \ldots\}$ is a set of messages;
5. $\Sigma \subseteq \{ch?, ch!\} \times M \cup Act$ is the set of events. Act is normal events. $\{ch?, ch!\} \times M$ is the synchronizing events.
6. $Init \subseteq Q \times X \times P(F)$ is a set of initial states;
7. $D(\cdot) : Q \rightarrow P(X)$ is a domain;
8. $E \subseteq Q \times Q$ is a set of edges;
9. $G(\cdot) : E \rightarrow L \times \Sigma \times P(X)$ is a guard condition;
10. $R(\cdot, \cdot) : E \times X \rightarrow P(X)$ is a reset map for a jump.
11. $F : Q \rightarrow P(Q \times X \rightarrow R^n)$ is the set of all vector fields for each state that satisfy all locations and guards of edges starting from this state (where $f(\cdot, \cdot) : Q \times X \rightarrow R^n$ is a vector field). For any $q_1 \in Q$, $F(q_1) = \{f \mid (\forall (q_1, q_2) \in E)(\forall (l, a, g) \in G((q_1, q_2)))(\exists t \in \mathbb{R}^+).(\dot{x} = f(q_1, x) \wedge x(t) \in l \cap g\};$

The different from the definition of hybrid automata is that we define a set of differential equations for each state that satisfy all guards of each edge from this state. We also add location set L as a field of locations. A location can be any physical locations like $\mathbb{R}$, $\mathbb{R}^2$, etc.

We say a differential equation $\dot{x} = f(q, x)$ satisfies a guard g, for example, we mean that there exists a time point t such that $x(t) \in g$. For each state q, there exists a set of differential equations $F(q)$ that satisfy all its guard conditions ($F(q)$ might be $\emptyset$ if no guards can be satisfied). This shows that in our spatial-hybrid automata we do not care the exact differential equation that trigger each edge. We do not care how to achieve the edge conditions, but the edge conditions itself.

4.4 Semantics of Spatial Hybrid Automata

Like hybrid automata, we give the transition semantics of spatial hybrid automata. The transition relation of spatial hybrid automata, as hybrid automata, constitutes of the labelled transition relation and the continuous transition relation.

Labelled Transition Relation. A triple $\rightarrow \subseteq St \times \Sigma \times St$ is called a labelled transition relation. We write $s \xrightarrow{a} s'$ if $(s, a, s') \in \rightarrow$.

Continuous Transition Relation. A triple $\Rightarrow \subseteq St \times \mathbb{R}^+ \times St$ is called a continuous transition relation. We write $s \xRightarrow{\delta} s'$ if $(s, \delta, s') \in \Rightarrow$.

Definition 42 (Transition Semantics). Let $sHa = (Q, X, L, M, \Sigma, F, Init, D, E, G, R)$ be a spatial hybrid automata, the transition relation is given as follows:

  i. **State.** For any (q_0, x_0), $(q_0, x_0) \in St$ iff $x_0 \in D(q)$ and $F(q_0) = \{f \mid (\forall (q_0, q) \in E)(\forall (l, a, g) \in G((q_0, q)))(\exists t \in \mathbb{R}^+).(\dot{x} = f(q_0, x) \wedge x(t) \in l \cap g\}$.

ii. **Labelled Trans.** For any $s = (q_0, x_0)$, $s' = (q_1, x_1) \in St$, $a \in \Sigma$, $s \xrightarrow{a} s'$ iff $(q_0, q_1) \in E$, $(x_0, a, x_0) \in G((q_0, q_1))$ and $x_1 \in R((q_0, q_1), x_0)$.

iii. **Continuous Trans.** For any $s = (q, x_0), s' = (q, x_1) \in St$, $\delta \in \mathbb{R}^+$ and $f \in F(q)$, $s \xRightarrow{\delta} s'$ iff $x_1 = x_0 + \int_{t_0}^{t_0+\delta} f(q,x)dt$ for some t_0.

The spatial hybrid automata allow all continuous transitions with the differential equations ranging in set $F(q)$ for each state q. While the definition of state and labelled transitions remain the same as hybrid automata.

4.5   Formal Model of $Data'_{TS}$

Given the data $Data'_{TP}$ in Section 4.1, we can build a spatial-hybrid automata sH_{aTP} shown in Fig. 8.

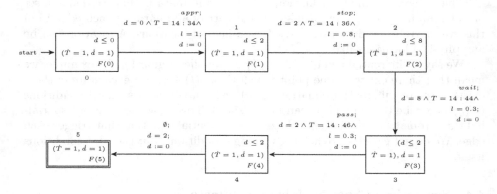

Fig. 8. Spatial-hybrid automata for $Data'_{TP}$

$sH_{aTP} = (Q, X, F, L, M, \Sigma, f, Init, D, E, G, R)$, where

- $Q = \{0, 1, 2, 3, 4, 5\}$;
- $\mathbf{X} = \mathbb{R}^3$, since there are 3 variables: T, d, l;
- $L = \{l = 1, l = 0.8, l = 0.3\}$ are the location conditions;
- $M = \emptyset$;
- $\Sigma = \{appr, stop, wait, pass\}$ is the set of alphabets;
- $Init = \{(0, x)\}$, where $x = [T = 14 : 34, d = 0, l = 0]$;
- D is the domain of invariants. e.g, $D(0) = (d \leq 0)$, $D(2) = (d \leq 8)$;
- $E = \{(0, 1), (1, 2), (2, 3), (3, 4), (4, 5)\}$ is the set of edges;
- G is guard condition. e.g, $G((2, 3)) = (l = 0.3, wait, d = 8 \wedge T = 14 : 44)$;
- R is a reset map. e.g, $R((2, 3), d) = 0$;
- F is the set of vector fields for each state. e.g. $F(0) = F_0 = \{f \mid (\exists t \in \mathbb{R}^+).(\dot{T} = f(0, T) = 1 \wedge \dot{d} = f(0, d) = 1 \wedge \dot{l} = f(0, l) \wedge d(t) = 0 \wedge T(t) = 14 : 34 \wedge l(t) = 1)\}$;

Compared with the automata of $Data_{TP}$ (Fig. 6), in sH_{aTP}, each variable is not only driven by one differential equation in each state, but a set of differential equations. Each edge is triggered by not only guards but also locations (L).

4.6   Condition-Triggered Transition System

In Section 4.3 we propose a 'location triggered' hybrid automata, it is an abstract automata where in each state, a set of differential equations (not only one) satisfy all guards and locations of edges out of this state. The differential equations that satisfy the guards are depending on the domain of 'locations', just as the $Data_{TS}$ and $Data'_{TS}$ shown above. More generally, we can define an abstract transition system called 'condition-triggered transition system', where each edge is triggered by one or several conditions with specific domains. See the next definition:

Definition 43. *A condition-triggered transition system $ctTS$ is a collection* $(Q, X, C, \Sigma, F, Init, D, E, G, R)$, $(q, x) \in Q \times X$ *as the state of $ctTS$, where*

1. *$Q = \{q_1, q_2, \ldots\}$ is a set of discrete states;*
2. *$X = \mathbf{R}^n$ is a set of continuous states;*
3. *$C = \{c_1, c_2, \ldots\}$ is a finite set of conditions;*
4. *Σ is set of alphabets.*
5. *$Init \subseteq Q \times X \times P(F)$ is a set of initial states;*
6. *$Dom(\cdot) : Q \rightarrow P(X)$ is a domain; ($P(X)$ denotes the power set of X)*
7. *$E \subseteq Q \times Q$ is a set of edges;*
8. *$G(\cdot) : E \rightarrow C \times \Sigma \times P(X)$ is a guard condition;*
9. *$R(\cdot, \cdot) : E \times X \rightarrow P(X)$ is a reset map for a jump.*
10. *$F : Q \rightarrow P(Q \times X \rightarrow R^n)$ is the set of all differential equations that satisfy all guards and conditions in C. For any state $q_1 \in Q$, $F(q_1) = \{f \mid (\forall(q_1, q_2) \in E)(\forall(c, a, g) \in G((q_1, q_2)))(\exists t \in \mathbb{R}^+).(\dot{x} = f(q_1, x) \wedge x(t) \in c \cap g\};$*

It is almost like a hybrid automata except that we take each condition c in C as a variable, it determines the set of differential equations $F(q)$ for each state q. We omit the discussion about it.

5   Conclusion

In this paper, we are trying to give a new developing method for software engineering by which somehow to solve the crucial problems existed in the field of nowadays software engineering. We mainly propose a data-driven framework and give some examples to explain our idea. We also give a case study of building the formal model of a special type of systems—spatial-temporal systems using our proposed framework.

The core idea of the data-driven framework that distinguishes it from other methods is that we want to design models through the data, since data is the most intuitive for human beings and somehow easy to acquire. In this paper we only show some example of how data and their relative models (might) look like, and choose a formal model for a special type of data. We have no idea about the way of generating model from data. It is a complex problem, and various from one another according to different types of data, different user requirements and some other factors. For the future work, maybe we will focus on building some concrete systems (under well-defined circumstances and clear users' requirements) and trying to find concrete algorithms to deal with the generating procedure.

Acknowledgment. This work is supported by the National Basic Research Program of China (No. 2011CB302802), the National Natural Science Foundation of China (No.61370100 and No. 61321064), Shanghai Knowledge Service Platform Project (No.ZF1213) and Shanghai Municipal Science and Technology Commission Project (No.14511100400).

References

1. Rumbaugh, J., Jacobson, I., Booch, G.: Unified Modeling Language Reference Manual, 2nd edn. Pearson Higher Education (2004)
2. Data-Driven: https://en.wikipedia.org/wiki/Data-driven_programming
3. Wirfs-Brock, R., Wilkerson, B.: Object-oriented design: a responsibility-driven approach. In: The Proceeding OOPSLA 1989 Conference Proceedings on Object-oriented Programming Systems, Languages and Applications, pp. 71–75. ACM, New York (1989)
4. Widom, J., Ceri, S. (eds.): Active Database Systems: Triggers and Rules for Advanced Database Processing. Morgan Kaufmann, San Francisco (1994)
5. Paton, N.W. (ed.): Active Rules in Database Systems. Springer, New York (1999)
6. Helmer, S., Poulovassilis, A., Xhafa, F.: Reasoning in Event-Based Distributed Systems. Springer, Heidelberg (2011)
7. Roussos, G.: Networked RFID: Systems, Software and Services. Springer, London (2008)
8. Samek, M.: State Machines for Event-Driven Systems (2003). http://www.barr-group.com/Embedded-Systems/How-To/State-Machines-Event-Driven-Systems
9. Nuseibeh, B., Easterbrook, S.: Requirements engineering: a roadmap (PDF). In: Proceedings of the Conference on the Future of Software Engineering, ICSE 2000, pp. 35–46 (2000). doi:10.1145/336512.336523
10. Chemuturi, M.: Requirements Engineering and Management for Software Development Projects (2013). doi:10.1007/978-1-4614-5377-2. ISBN: 978-1-4614-5376-5
11. Kotonya, G., Sommerville, I.: Requirements Engineering: Processes and Techniques. John Wiley & Sons, September 1998. ISBN 0-471-97208-8
12. Thayer, R.H., Dorfman, M. (eds.): Software Requirements Engineering, 2nd edn. IEEE Computer Society Press (1997). ISBN 0-8186-7738-4
13. Mu, K., Kiu, W., Jin, Z., Lu, R., Yue, A., Bell, D.: Handing Incosistency in Distributed Software Requirements Specification Based on Prioritized Merging. Fundamenta Informaticae **91**(3–4), 631–670 (2009)
14. Chen, Y.: Stec: A location-triggered specification language for real-time systems. In: ISORC Workshops, pp. 1–6. IEEE (2012)
15. Stutz, M.: Get started with GAWK: AWK language fundamentals. developerWorks. IBM (retrieved October 23 2010)
16. Hopcroft, J.E., Motwani, R., Ullman, J.D.: Introduction to Automata Theory, Languages, and Computation, 3rd edn. Addison-Wesley Longman Publishing Co., Inc., Boston (2006)

17. Rabin, M.O., Scott, D.: Finite automata and their decision problems. IBM J. Res. Dev. **3**(2), 114–125 (1959)
18. Alur, R., Dill, D.L.: A theory of timed automata. Theor. Comput. Sci. **126**(2), 183–235 (1994)
19. Asarin, E., Caspi, P., Maler, O.: A kleene theorem for timed automata. In: LICS. IEEE Computer Society, pp. 160–171 (1997)
20. Alur, R., Courcoubetis, C., Henzinger, T.A., Ho, P.-H.: Hybrid automata: an algorithmic approach to the specification and verification of hybrid systems. In: Hybrid Systems, pp. 209–229. Springer, London (1993)

Requirements Acquisition via Crowdsourcing

A Recommendation Algorithm for Collaborative Conceptual Modeling Based on Co-occurrence Graph

Kai Fu[1,2], Shijun Wang[1,2], Haiyan Zhao[1,2(✉)], and Wei Zhang[1,2]

[1] Institute of Software, School of EECS, Peking University, Beijing, China
{fkai1993,wangshijun,zhhy.sei,zhangw.sei}@pku.edu.cn
[2] Key Laboratory of High Confidence Software Technology,
Ministry of Education of China, Beijing, China

Abstract. Conceptual models are models used to describe objects or systems in the real world. The quality of a conceptual model heavily depends on the domain knowledge and modeling experience of the individual modeler. Collaborative conceptual modeling is an effective way of building models by taking advantage of collective intelligence. This paper proposes a Co-occurrence Graph based Recommendation Algorithm (CGRA) to implement the collaborative mechanism of conceptual modeling systems. CGRA, inspired by association rule mining algorithm, is an incremental data updating algorithm. The computational complexity of CGRA is much lower than that of the traditional association rule mining based algorithms, while the recommendation effectiveness of these two are almost the same in our collaborative conceptual modeling system, which is revealed by the experiments we have conducted.

Keywords: Collaborative Conceptual Modeling · Recommendation · Association Rule Mining · Co-occurrence Graph

1   Introduction

Conceptual models are models used to describe objects or systems in the real world. For most areas of engineering, to build the corresponding conceptual model on the stage of requirement analysis is of great importance. However, the quality of a conceptual model heavily depends on the knowledge and experience of the individuals who build it. It is scarecely feasible to build a good domain specific conceptual model all by one person, due to the limited kowledge of an individual. Traditionally, conceptual modeling requires a dozen of experts who are familiar with the corresponding domain knowledge to gather together and reach a consensus. Top-level experts are not always available for every organization, and the whole procedure of experts' meeting is time consuming.

Fortunately, in the Internet age, we could utilize collective intelligence to compensate for the lack of individual knowledge and top-level expert modelers.

K. Fu and S. Wang—These authors contributed equally to this work and should be considered co-first authors.

L. Liu and M. Aoyama (Eds.): APRES 2015, CCIS 558, pp. 51–63, 2015.
DOI: 10.1007/978-3-662-48634-4_4

Collaborative conceptual modeling system provides a way to make use of collective intelligence [7]. In a collaborative conceptual modeling system, people could build their own conceptual models while enjoying the benefit of collective intelligence from recommendation information pushed by the modeling system. Individuals use their conceptual modeling domain knowledge independently, while collaborative conceptual model will recommend some concept elements with high quality in collective conceptual model to the individual in the process of modeling, to help individuals to explore and establish a better conceptual model. As the individual conceptual models getting better, the quality of the collaborative conceptual model is promoted in return, which forms a positive feedback loop.

Recommendation plays a key role in such collaborative conceptual modeling systems. During the modelling process, the communication among individuals depends mainly on the information from the recommender system as an indirect interaction. So the results of the collective conceptual model for the individuals depends on the recommender system. If the recommendation results do inspire the individuals, a positive feedback loop is established, which further enables the collective conceptual model evolve eventually to a conensual conceptual model by the majority of individuals.

Traditional recommendation approaches include collaborative filtering (CF) based approaches, content based approaches and knowledge based approaches [2]. These approaches are more applicable to the scenarios like online shopping item recommendation, music recommendation and movie recommendation, in which, recommender systems tend to recommend some items that have similar features to those that user has purchased before according to the user's need. However, due to the characteristics of conceptual modeling, most of those approaches are difficult to recommend a good concept element to the individual except for the association rule mining based CF approaches [1][6], since each concept in the individual conceptual model is unique, any concept that the individual currently does not possess would not be similar to those already in the individual conceptual model.

On account of the characteristic of association rule mining based CF approaches, they could be well applied in collaborative conceptual modeling systems. However, there still are some shortcomings to use these approaches in our conceptual modeling system. On one hand, they tend to get too many candidate item sets and therefore lead to high I/O overhead. On the other hand, the computational complexity of those approaches is rather high. In our collaborative conceptual modeling system, the running time of the recommendation algorithm must be taken into account to guarantee real-time reponses to the individual operations.

In this paper, we propose a Co-occurrence Graph based Recommendation Algorithm (CGRA), a simplfied association rule mining based algorithm, to solve the problems mentioned above. The computational complexity of CGRA is much lower than that of the traditional association rule mining based algorithms, while the recommendation effectiveness of these two are almost the same in our collaborative conceptual modeling system.

The term co-occurrence graph and co-occurrence matrix have been widely used in the literature of recommender systems and text indexing [4], usually denoting the relationships between items and users, or those between documents and words. On the contrary, co-occurrence graph in our work has a different meaning: The two vertices of a edge   are both conceptual elements.

The remainder of this paper is organized as follows. Section 2 introduces some preliminaries aboutcollective conceptual model and association rule mining. Section 3 presents the basics about our CGRA, including the definition of Co-occurrence Graph (CG, the core data structure we use in our algorithm), the CG building method, our recommendation stragegies and finally a incremental CG updating algorithm. Section 4 discusses the advantages of CGRA over traditional association rule mining based algorithms in a collaborative conceptual modeling system, both theoretically and experimentally. Finally, section 5 concludes our work.

2 Preliminaries

CGRA is a modified version of association rule mining which is based CF approaches, to make the traditional association rule mining recommendation appliable to the scenarios during collaboritive conceptual modeling. Co-occurrence graph is uesd to describe and store a certain kind of association rules. The co-occurrence graph in CGRA is constructed from collective conceptual model, a fusion version of each individual's model. In this section, we first focus on the collective conceptual model, and then review the traditional algorithms and stratergies of association rule mining approaches used in recommender system.

2.1 Collective Conceptual Model

A conceptual model could be represented as a class diagram, with concepts as classes with attributes. Under collaborative scenarios, the collective conceptual models should be capable of reflecting the viewpoints of a group. However, the traditional class diagram representation is hard to describe the characteristic of group. We therefor have to make some changes for the collective conceptual model. For instance, the traditional class diagram model is a hierarchical structure, the attributes are part of the class, and the classes rely on relationship to connect each other. It's difficult to depict the collective class diagram model, owning to the relationship between attributes and classes. In this paper, the hierarchical structure of collective class diagram model is converted into a model of collective class diagram based on graph structure. In this graph, there are three primary types of nodes: concept nodes, relationship nodes and value nodes. Each node attaches a counter recording the corresponding number of users who reference it. Each concept node represents one class. Each relationship node represents the relationship of two connected concept nodes. Each value node represents a value. And each edge has a name. Figure 1 shows a structrual graph representaion (right part) and its corresponding conceptual model(left part).

In addition, we refer to the behavior user create an element or reference the element created by other user as a reference.

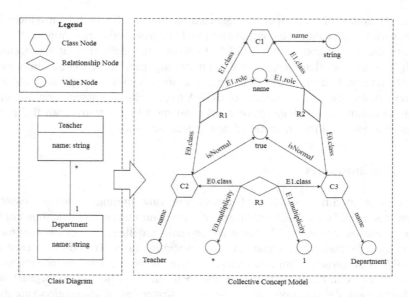

Fig. 1. The conversion of a conceptual model from class diagram representation to its structrual graph representaion

2.2   Association Rule Mining Based CF Approaches

Association rule mining are used to discover interesting relationships hidden in large data sets [1]. To better formulate the problem, we first give some definitions.

$I = \{i_1, i_2, ..., i_d\}$ denotes the set of all items in a market basket data. $T = \{t_1, t_2, ..., t_N\}$ denotes the set of all transactions. An itemset is a collection of zero or more items, i.e. a subset of I. K-itemset means an itemset contains k items. A transaction contains a subset of items chosen from I. The support count of $\sigma(X) = |\{t_i | X \subseteq t_i, \ t_i \in T\}|$.

Association rule is represented as $X \rightarrow Y$, where X and Y are disjoint itemsets, i.e. $X \cap Y = \emptyset$.

There are two primary metrics to measure the quality of a given association rule $X \rightarrow Y$.

One is the support $s(X \rightarrow Y) = \frac{\sigma(X \cup Y)}{N}$, which measures the degree of the correlation between itemsets.

The other is the confidence $(X \rightarrow Y) = \frac{\sigma(X \cup Y)}{\sigma(X)}$, which measures the significance of the correlation between itemsets.

Using these definitions, the association rule mining problem could be formulated as follows: *Given a set of transactions T, find all the rules having support ≥ minsup and confidence ≥ minconf, where minsup and minconf are the corresponding support and confidence thresholds.*

The basic way to solve this problem is to divide it into two major steps, frequent itemset generation and rule generation.

Frequent itemset generation would find all the itemsets that satisfy the *minsup* threshold (frequent itemsets). And during rule generation step, it extracts all the high-confidence rules from the frequent itemsets found in the previous step (strong rules).

To lower the high computational complexity of the solution, the association rule mining problem in traditional recommender system [2] is a little bit different:

Given a set of transactions T, find all the k-itemset to 1-itemset rules having $|\{rule|1\text{-}itemset_i$ *as the head of rule}*$| \in [minNum$, $maxNum]$, *where i=1,2,...,d, and confidence* $\geq minconf$, *where minconf is the corresponding confidence thresholds.*

Note that the support threshold is not set in advance, which prevent the situation that too many or too few rules are generated for a certain item when the traditional association rule mining is applied.

3   The Co-occurrence Graph Based Recommendation Algorithm

The recommendation scenarios in colloboritive conceptual modeling are sort of different from those in the traditional recommender systems. In traditional recommendations, the quality of a recommendation result is measured by whether the recommended items are adopted by the user. Whereas, the recommendation in a conceptual modeling focuses on whether it could inspire the individual modeler to build a better conceptual model at last. Therefore, the the precis ion of one single recommendation is not of that great significance. On the other hand, the recommendation requests would be sent in each step during one individual's modeling activity, requiring an even stricter time constraint compared to traditional recommender systems.

To reach a solution, we further improved the association rule mining based recommendation algorithm. It has much lower computational complexity while maintaining the recommendation quality required by colloboritive conceptual modeling systems. What's more, we promote an incremental recommendation algorithm that could be applied to a real colloboritive conceptual modeling system. In the end, we analysis the effect and advantage of our algorithm compared with the traditional association rules.

3.1   Co-occurrence Graph

To lower the complextiy of association rule mining algorithm, we only focus on the 1-itemset to 1-itemset rules. Under such circumstances, the rules could be represented as a graph, whose vertices are concepts or relationships in the modeling system and edges are the association rules. Such a graph is called a co-occurrence graph, suggesting that the edges are co-occurrence rules of nodes in the graph.

Thus, our problem could be formulated as follows: *Given a set of transactions T, calculate the confidence of all the 1-itemset to 1-itemset rules.*

The transactions are mapped to individual users, and the items to concepts and relationhips between concepts.

In the rest of this paper, we would adopt the following definitions to describe a co-occurrenc graph as is shown in Figure 2. Element node $e_i, where\ i = 1,2,...,n$

denotes a concept node or relationship node in the co-ocurrence graph. Element set $E_{set} = \{e_1, e_2, ..., e_n\}$ denotes a collection of element nodes. User $u_i, where\ i = 1, 2, ..., n$ denotes an individual modeler in the collaborative conceptual modeling system. User set $U_i = \{u_1, u_2, ..., u_n\}$ denotes a collection of users that reference a certain element node.

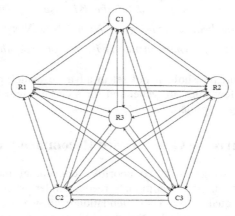

Fig. 2. A co-occurrence graph

3.2 The CG Building Method

For each element e_i, it has a set U_i containing all the users who reference the node. For all the nodes j except the node i in E_{set}, if the user has referenced the element node i, then the probability that he will reference the node j is $p(j|i) = \frac{|U_i \cap U_j|}{|U_i|}$, in which the element node j belongs to the set E_{set} and $i \neq j$.

Then our algorithm could be described as follows.

1) Traverse an element node from the node set E_{set}. If the traverse ends, go to step 7; else select the element node i.
2) For element node i, get the reference user set U_i.
3) Traverse an element node from the node set E_{set} except element node i. If the traverse ends, go to step 1; else select element node j, go to step 4.
4) For element node j, get the reference user set U_j.
5) Build an edge from node i to node j, and set the edge value $P(j|i) = \frac{|U_i \cap U_j|}{|U_i|}$.
6) go to the step 3.
7) End.

3.3 Recommendation Strategies

There are three main recommendation strategies. They respectively corresponds to three different recommendation scenarios. We will first introduce the three recommendation scenarios before we talk about the corresponding recommendation strategies.

The first recommendation scenario is for new users who do not reference any model element, in which case we could not get any information from the new users' behaviors. The second recommendation scenario is that the user has created some elements in his conceptual model, where the user need some recommended elements to promote his understand of the collective conceptual model. The third scenario is the most common scenario in our recommender system: users select one of the elements that he has already referenced.

Strategy One: Directly recommend some elements that have high reference count in our collective model for new users

Strategy Two: Multiply the correlation intensity between the elements user referenced and other elements by support to get a result. Sort these results and get recommendation order of the elements that are not referenced by user.

Strategy Three: Suppose that the user U_1 references the element A. And we get the first m elements which have the largest P value frsom the nodes that are not referenced by U_1 in the co-occurrence graph. Then we sort these nodes by P value and put the first m element nodes in our collective conceptual model. Finally we can get the corresponding concepts of these elements and recommend these concepts to the user.

The process of choosing strategies above is based on the co-occurrence graph. Although the co-occurrence graph has the most significant effect on strategy 3, all the three strategies are based on the related properties of the co-occurrence graph in order to get the results of recommendation as soon as possible. Therefore, the co-occurrence graph has a great influence on the results of recommendation.

3.4   Incremental CG Updating Algorithm

To further reduce the time consumption of our current algorithm, an incremental updating algorithm could be introduced. With the help of this incremental approach, our algorithm could generate a new co-occurrence graph faster when the collective concept model is changed, resulting in more accurate concepts being recommended to users.

Figure 3 is a co-occurrence graph made of three nodes, A, B and C. The corresponding user sets of the three nodes are U_A, U_B, U_C. If the state that user u_i has never referenced A changes to the state that the user u_i references A, the co-occurrence graph must be updated. At the same time, the state of node A in the graph is changed (The number of users that reference node A increases). And all the edges starting from A and all the edges ending in node A are influenced, which means the P value of these edges are changed.

If we updates all the influenced edges, 2n edges need to be updated. However, we notice that if the user u_i has referenced the node A, then we don't need to recommend A to u_i. Thus it has no influence to u_i whether or not updating the edge ending in node A, For u_i, only the edges that start from the node A have influence on the results of recommendation. In addition, if more than one user modify node A in a certain time, the modifying of edges that ending in node A during this time is meaningless. Thus, this recommendation this time should not include node A, no matter how to modify the edges.

To solve this problem, we take the method namely delay modifying. That means that after node A is modified by user u_i, we only need to update all the edges that start from A. And we mark node A that it needs to update all the edges that end at it. We will not update all the edges that end at node A until we need to recommend node A to those users who have never referenced node A. In this way, we need to update only n edges no matter how many users are modifying the node A, because the n edges that end at node A are updated until we need to recommend node A to a user that has never referenced node A.

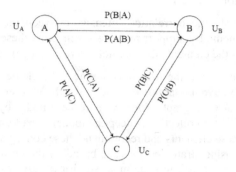

Fig. 3. Calculations in a co-ocurrence graph

The specific algorithm is shown as follows:

1) The user u_i request the recommender system to recommend concepts to him.
2) If the user u_i modify the node A, we will update all the edges that end in node A: $P(B|A) = \frac{|U_A \cap U_B|}{|U_A|}$, in which B belongs to E_{set}. And we value the Flag[A] true: Flag[A]=true. If not, go to step 3.
3) Go through the data records in the array Flag and determine if the nodes in array Flag except node A have been referenced by user u_i. If all of them have been referenced, go to step 7. If not, go to step 4.
4) Get the node X that is marked true and is not referenced by user u_i in array Flag and mark the node X false: Flag[X]=false.
5) Update all the edges that end in node X: $P(X|B) = \frac{|U_X \cap U_B|}{|U_B|}$, in which B belongs to E_{set}.
6) Go through the data records in the array Flag and determine if there exist a node in all the nodes in array Flag except node A that is marked true and has not been referenced by user u_i. If there exists such a node, go to step 4. If not, go to step 7.
7) Use the updated co-occurrence graph to fit the three scenarios in 3.3. And recommend the corresponding element nodes to the user u_i (Node A won't be recommended in this round). The final recommendation results are generated.
8) End.

4 Algorithm Analysis and Experimental Evaluations

In this section, we will first explain why CGRA in this paper is a more appropriate choice to be used in collaborative modeling system compared with the traditional association rule mining algorithms. Then we will prove the efficiency of our algorithm by comparing the experimental results of our algorithm with a traditional association rule mining based algorithm.

4.1 Algorithm Analysis

As is described in section 3.3, there are three recommendation scenarios in collective collaborative modeling system. For the first scenario, it is reasonable to recommend the most often referenced elements in the current collective model. For the third scenario, recommendations of each element need to be provided to the user. In this case, it's time consuming using traditional association rule algorithms, in which 1-to-1 association rules are produced at the same time with m-to-n rules. However, CGRA in this paper will produce 1-to-1 association rules quickly and recommend the model most relevant elements to the user. It reduces a lot of time compared to the traditional association rule algorithm. For the second scenario, our algorithm only uses 1-to-1 association rules not considering m-to-n association rules. In the following, we will prove that it is reasonable.

It is important to recommend items that could attract a user for online stores, movie and music websites for profit. So it is necessary and essential to recommend a user items most relevant to that those user has bought.

However, we aim to help a user build and refine his conceptual model in collaborative modeling. During the process, on one hand, we hope that the user can obtain some helpful elements from the recommender system in our collective conceptual model. On the other hand, the user should not completely depend on the recommend system to build a model. Instead, the user needs these recommended elements to expand thinking to build a more perfect model. In addition, it makes little difference when recommending an element since modeling is a continuous process. As a result, the recommender system using collaborative modeling in this paper doesn't always recommend the most relevant elements to the user in first time.

It doesn't mean that the recommender system recommends random elements to the user. We still recommend rather helpful elements to the user. The order of the recommended elements is not strict in our algorithm.

Although m-to-n association rules could produce related elements earlier compared to 1-to-1 association rules, it is time consuming for huge amount of data to compute the m-to-n association rules due to the exponential computational complexity of this problem [5]. However, due to the fact that a user depends on the recommender system to interact with collective conceptual model during the process of modeling, we hope that the recommender system is capable of adjusting itself quickly to the collective conceptual model.

Considering the fact mentioned above that the order of elements recommended doesn't matter to a collective collaborative modeling system and the m-to-n association rules are really time consuming, we only consider the 1-to-1 association rule in our algorithm and discard m-to-n association rules to achieve a polynomial computational complexity. The effectiveness of our algorithm will be validated in the experimental evaluation parts of this section.

4.2 Experimental Evaluations

All of the experiments in this paper are based on data for collaborative conceptual modeling due to that the recommender system in this paper is specifically for collaborative conceptual modeling. The data in this paper come from a collective conceptual model named Course Management System created by 15 users, which contains 67 class nodes and 360 relation nodes. We compare the traditional association rule algorithm and our CGRA based on co-occurrence graph. We choose the efficient algorithm FP-Growth improved from the traditional association rule mining algorithm Apriori as the represent of association rule algorithm [3]. FP-Growth algorithm produce both 1-to-1 association rules and m-to-n association rules.

In the third scenario mentioned in 3.3 section, the parameter support and the parameter confidence are both set as zero in FP-Growth algorithm to compute the correlation intention of elements that have been used in model. It takes more than 12 hours to obtain the results. However, it takes only 3 seconds in our algorithm. In addition, the recommendation results of the two algorithms are the same since only 1-to-1 association rules are used in this scenario.

Thus, the results of the two algorithms in the second scenario are compared in the following.

In the Course Management System collective model, two users (user 1 and user 2) participate in the model. User 1 completely adopts recommended elements and does not create any new element by himself. User 2 chooses some elements from the recommended elements, think over and build his own model.

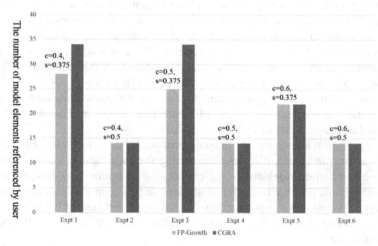

Fig. 4. Recommendation results of user 1 using the two algorithms respectively (*c* for *confidence*, *s* for *support*)

To make the recommendation results more convincing, the parameters support and confidence are set as different values in FP-Growth algorithm to get rid of some unconvincing association rules. We will compare a group of results of CGRA and FP-Growth with different values of support and confidence.

The specific process of the experiment is shown as follows.

In the initial state, the recommender system will recommend the top-four most referenced elements to a user due to the lack of individual modeling information. In the following time, the recommender system will recommend 6 elements at most every time until the user stop the process of modeling.

Figure 4 shows the recommendation results of user 1 using the two algorithms respectively. For FP-Growth algorithm, each bar represents the number of referenced elements at a certain value of confidence and support. For CGRA, each bar represents the number of referenced elements that contains all the elements recommend by FP-Growth. As what the figure 4 shows, our recommendation results cover those in FP-Growth, which means that our algorithm is more effective. In fact, the order of the recommended elements is almost the same except for a few elements. However, it does not influence the final model results.

Due to the small scale of this data set, there is no significant difference between our algorithm and FP-Growth algorithm in time consuming in this experiment. As we all know, it takes a long time for a large amount of data using FP-Growth algorithm. Besides, that our algorithm is faster could be inferred from the behavior of FP-Growth algorithm in scenario 3. The corresponding experimental results are shown in Table 1.

Table 1. Run time evalution

Algorithms	Average running time (s)
FP-Growth (c=0.6, s=0.375)	4
FP-Growth (c=0.6, s=0.5)	3
FP-Growth (c=0.5, s=0.375)	4
FP-Growth (c=0.5, s=0.5)	3
FP-Growth (c=0.4, s=0.375)	4
FP-Growth (c=0.4, s=0.5)	3
CGRA	3

What's more, we notice that when the confidence is set as 0.5, support is set as 0.375 or the confidence set as 0.4 and support set as 0.375, the recommendation result of our algorithm includes more elements than that of FP-Growth until all the elements is recommended by FP-Growth. To prove that our algorithm will recommend more and more helpful elements than FP-Growth algorithm in users' behaviors, we compare the results of our algorithm and the FP-Growth algorithm of user 2 with two groups of parameters. Figure 5 shows the results of the two algorithms of user 2.

In summary, the experiments above show that the recommendation effectiveness of our algorithm is not worse than that of the traditional association rule mining algorithm. However, due to the simple structure of our algorithm, based on co-occurrence graph, the algorithm in this paper is much faster than the traditional association rule mining algorithm. Of course, because the collaborative conceptual modeling tool is still in its experimental stage, the amount of participants is not large. In the future work we will verify the experimental results with a larger number of participants and a more complex modeling environment.

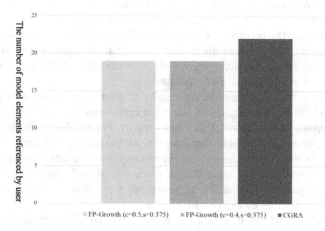

Fig. 5. The results of the two algorithms of user 2

5   Conclusions

The computational complexity of our co-occurrence graph based recommendation algorithm is much lower than that of the traitional association rule mining based algorithms, while the recommendation effectiveness of these two are almost the same in our collaborative conceptual modeling system.

Acknowledgements. This research was supported by the National Basic Re- search Program of China (the 973 Program) under grant 2015CB352201; Science Fund for Creative Research Groups of the National Natural Science Foundation of China (Grant No.61421091); and the National Natural Science Foundation of China under grants 91318301 and 61272163.

References

1. Agrawal, R., Imieliński, T., Swami, A.: Mining association rules between sets of items in large databases. ACM SIGMOD Record, ACM. **22**(2), 207–216 (1993)
2. Bobadilla, J., Ortega, F., Hernando, A., et al.: Recommender systems survey. J. Knowledge-Based Systems. **46**, 109–132 (2013)
3. Han, J., Pei, J., Yin, Y.: Mining frequent patterns without candidate generation. ACM SIGMOD Record, ACM. **29**(2), 1–12 (2000)

4. Huang, Z., Chung, W., Chen, H.: A graph model for E-commerce recommender systems. J. Journal of the American Society for Information Science and Technology. **55**(3), 259–274 (2004)
5. Kosters, W.A., Pijls, W., Popova, V.: Complexity analysis of depth first and fp-growth implementations of apriori. In: Perner, P., Rosenfeld, A. (eds.) MLDM 2003. LNCS, vol. 2734, pp. 284–292. Springer, Heidelberg (2003)
6. Lin, W., Alvarez, S.A., Ruiz, C.: Efficient adaptive-support association rule mining for recommender systems. J. Data Mining and Knowledge Discovery. **6**(1), 83–105 (2002)
7. Zhang, W., Zhao, H., Jiang, Y., et al.: Stigmergy-Based Construction of Internetware Artifacts. J. Software, IEEE. **32**(1), 58–66 (2015)

Requirement Acquisition from Social Q&A Sites

Ming Xiao[✉], Gang Yin, Tao Wang, Cheng Yang, and Mengwen Chen

National Key Laboratory for Parallel and Distributed Processing,
College of Computer Science, National University of Defense Technology,
Changsha, Hunan Province, China
xiaoming-7@qq.com, {jack.nudt,taowang2005}@nudt.edu.cn

Abstract. Social Q&A sites have changed the way of knowledge sharing in software communities. Comparing to the traditional mail-list, bug/change repositories, software forums and software marketplaces, users and developers are more active in social Q&A sites, and social Q&A sites are more open and free. The feedbacks from users have much potential valuable information, such as feature requests, bugs or sentiment, but there also exists lots of noise especially for social Q&A sites. How to mine the useful information from the feedbacks in social Q&A sites has become a problem. This paper focuses on the feature requirements in requirement acquisition, which can be used to assist software development. We propose an effective approach, which combines Support Vector Machine (SVM) with requirement dictionary to find the questions about feature requests from the posts in social Q&A sites. We evaluate the approach on available dataset, and compare it to the other different approaches. The results show that the automatically requirement acquisition through improved SVM approach is useful and can significantly decreases the manual effort.

Keywords: Requirement acquisition · Feature requests · Q&A sites

1 Introduction

In software development, requirement acquisition is very important, and the software can be successful depends on whether it meets the need of users. So user feedbacks are valuable for software development. With the development of Internet technologies, more and more software developments rely on Internet communities to gain the feedbacks or to communicate with users. More and more users also join in the communities to involve in the software development process, especially provide valuable suggestions for software developments. For example, in App Store users deliver the feedback after they use an app, and the information is useful for developers because it can assist developers to know what users hope the app to be and what features should be added in next versions.

There are many kinds of software communities, such as software forums, mail-list, and software marketplace. These years have witnessed the rapid development of social Q&A sites because they can provide better environments for

© Springer-Verlag Berlin Heidelberg 2015
L. Liu and M. Aoyama (Eds.): APRES 2015, CCIS 558, pp. 64–74, 2015.
DOI: 10.1007/978-3-662-48634-4_5

users to express their views or exchange their opinions. As the biggest programming Q&A site, Stack Overflow (www.stackoverflow.com) has achieved a great success in these years and is gaining more and more focuses. This can be a good description of the advantages of social Q&A sites. Bogdan Vasilescu [1] finds that the users in mail-list are migrating to Stack Exchange, a network of 145 communities that are created and run by experts and enthusiasts who are passionate about a specific topic [2]. Stack Exchange builds libraries of high-quality questions and answers focusing on each community's area of expertise, such as Stack Overflow (www.stackoverflow.com) for programming problems, WordPress (wordpress.stackexchange.com) for WordPress development, Android Enthusiasts (android.stackexchange.com) for enthusiasts and power users of the Android operating system, Unix&Linux (unix.stackexchange.com) for users of Linux, FreeBSD and other Unix-like operating systems. So the feedbacks in these Q&A sites have large numbers of valuable information for software development, and the number of questions is huge and increases quickly. But the questions are about many topics such as bugs, feature requests or even complaints. In this paper we focus on feature requests which are new features or re-design of existing features, and we hope to extract the questions about feature requests from the whole questions in social Q&A sites to assist software development.

In short, this paper makes the following contributions:

- Firstly focus on the requirement acquisition for software development from social Q&A sites which have become the most active communities;
- Based on the previous research, implement the approach based on linguistic rules to extract the valuable questions about feature requests;
- Propose a requirement acquisition approach based on Support Vector Machine (SVM) and improve the process of feature selecting based on TF-IDF and requirement dictionary about feature requirements;
- Evaluate the approach based on SVM and the approach based on linguistic rules, and the results show that our approach can get better performance and decrease the manual effort significantly.

2   Related Work

It seems that there are many similar studies on requirement acquisition. C. Iacob and R. Harrison [3] designed a MARA (Mobile App Requirement Analyzing) prototype, which can be used to automatically index the mobile app feature requests from online views. They got a set of 237 linguistic rules and used them to refer the other reviews, and they ranked the reviews based on the frequency and length of feature requests. M. Harman, Y. Jia, and Y. Zhang [4] focused on feature requests information from the text of app store descriptions. L. V. Galvis Carreño and K. Winbladh [5] alleviated the process of developer getting information for the next version of software through automatic topics extraction.

To our best knowledge, the most similar work to ours is AR-Miner [6], which facilitated mobile app developers to discover the most "informative" user reviews from a large and rapidly increasing pool of user reviews. But our work focus on social Q&A sites and the questions posted by users to extract the valuable questions about feature requests to assistant the software development.

3   Approach

In this section, we describe our approach based on SVM and the approach based on linguistic rules in detail. Firstly, we introduce how we handle the natural language text from social Q&A sites. Then we introduce the SVM and how to identify the features in SVM. At last, we present the approach based on linguistic rules in detail.

3.1   Preprocessing

Natural language text is unstructured, and we have to preprocess the text to get a structured data that the computer can recognize and compute. The general method to preprocess the natural lounge is VSM (Vector Space Model), which can translates the natural language text into vector space. In our approach, we also process the text with VSM, and the steps are listed as follows:

step 1 Get the texts and delete the useless tags such as HTML tags. The only valuable information in our method is pure text that users have posted;

step 2 Lowercase the words, because we needn't consider the letter case;

step 3 Tokenize the text to words and these words will be used to build a vector to represent the text;

step 4 Remove the punctuations, which are useless, so we need to remove them automatically;

step 5 Remove the stop words, which often are used in English expression but contain little information, and there is no need to keep these words;

step 6 Stem the words to root words, because the words in text have different tenses, but we needn't consider the difference and should regard them as the same;

step 7 Remove the words that are used less than 3 times.

After upper processing we have mapped the natural language text to vector space, and the vector will be used to compute with the computer.

3.2   Constructing Classifier Based on SVM

Our core work is to construct a classification machine which can automatically classify the documents into two categories, of which one is about feature requests and the other is about non-feature requests. We choose Support Vector Machines (SVM), which is widely used in text classification and fits our problem. SVM is a supervised learning model in machine learning field. In supervised learning, we need the training data that consists of a set of training examples. A supervised learning algorithm analyzes the training data and produces an inferred function, which can be used for mapping new examples [7]. In our approach, we input the training data that consist of training examples labeled as feature (F) or non-feature (NF) to SVM to train a classifier. After that, when a new question is inputted in the trained SVM model, the model will return the result about whether this question is about feature request or not. Fig. 1 depicts the overview of SVM approach in detail.

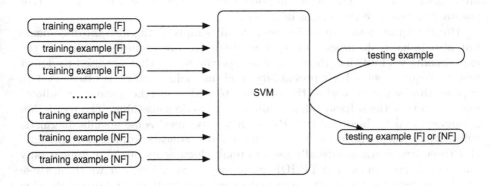

Fig. 1. Overview of SVM approach

The core idea of SVM is not complicated. SVM constructs a hyper-plane to separate the two kinds of training examples. In our experiment, we are lucky to use a successful and widely used SVM implementation named LIBSVM [8]. LIBSVM is a library for Support Vector Machines (SVMs) [9]. It is developed to help users to easily apply SVM to their applications by professor Chang from national Taiwan University.

3.3 Feature Selection in SVM

Focusing on the questions in social Q&A sites, we regard the question as a document. The weight of every word in a document is not the same. It is obvious that a word is used in a document frequently and is rarely used in other documents, and the weight of the word should be higher. On the contrary, a word is used in document rarely but is often used in other documents, and the weight of the word should be lower. So we should evaluate the importance of words and compute the weight of every word. Term Frequency-Inverse Document Frequency (TF-IDF) is widely used to compute the weight of words in documents. In which, Term Frequency (TF) means the frequency a word appears in a document and Inverse Document Frequency (IDF) means the reciprocal of document frequency. The specific definition of TF-IDF is as follow:

$$tf_{i,j} = \frac{n_{i,j}}{\sum_k n_{k,j}} \tag{1}$$

$$idf_i = log(\frac{|D|}{|j : t_i \in d_j|}) \tag{2}$$

$$tfidf_{i,j} = tf_{i,j} \times idf_j \tag{3}$$

Where $n_{i,j}$ means how many times a word i is used in document j, $|D|$ means the number of all the documents and $|j : t_i \in d_j|$ means how many documents are consist of word i. For word i in document j, we synthesize the TF and IDF together to evaluate the weight of a word.

But in requirement acquisition, we not only simply distinguish the documents but also judge a document whether is about feature requests or not. In the experiments we find that there are some specific words that are used to depict feature requests with high probability, such as "add", which is often used to express that someone wishes the software to add a specific feature or "allow" may mean that the software should allow users to do something. The regulation is obvious and C. Iacob and R. Harrison [3] also used some specific linguistic rules, which users often use, to find the feature requests from user comments. But these key words are usually used in many documents but may appear only once in a document, so the TF-IDF value may be very low, but we think these words can depict a document about feature requests well. So we borrow the idea from MARA [3] and build a requirement dictionary, which contains the words that are usually used to express feature requests. The partial key words which are often used to represent feature requests are listed in Table 1.

Table 1. Keywords about Feature Requests

add, allow, complaint, could, hope, improvement, instead, lacks, maybe, missing, must, need, please, prefer, request, should, suggest, want, will, wish, would, tag, allow, hope, improve, want, auto, let, filter...

We want to increase the weight of these key words because they can represent the documents we want well. In our approach, we simply duplicate these words in a document three times. For example, if "add" is used in a document, we extend the frequency of the word three times and there will be three "add" in the document. After the process, the IDF value will not change but the TF value can increase obviously.

3.4   Feature Acquisition Based on Linguistic Rules

To compare with the SVM approach, we implement a prototype based on MARA (Mobile App Review Analyzer) [3] that was proposed to retrieve and analyze mobile app feature requests from online reviews through a set of linguistic rules. We want to know whether the approach based on linguistic rules can fit the problem of our experiment or not. In our experiment, we choose 200 questions which are tagged "feature-request" by users to define a set of linguistic rules, and some illustrative examples are depicted in Table 2. In our experiment of linguistic rules, we only consider the titles of the questions because of the complexity of the bodies of the questions. After defining the linguistic rules which are usually used to express feature requests, we use these linguistic rules to judge whether the new question is about feature requests or not. The manual defining of the linguistic rules is time consuming and labor intensive. At last we define a set of 32 linguistic rules which are usually used to express feature requests.

Table 2. Linguistic Rules Elicited from 200 Training Examples

Linguistic rule	Example contexts
add/change ... to ...	add ability to cancel flags; change color of inline links in SO;
let's/let us ...	Let us bring an end to the "robo-reviewer" war: Phase 1-2; Let's improve the How to Ask page(s);
filter/improve ...	Filter 10k tools by a tag; Improve the "404 Not Found" error page for deleted questions;
can/should we ...	Can we have a delete review queue?; Should we be able to flag suggested edits?
sth. should ...	"Similar Questions" search should take the tags into account; Room owners should be allowed to accept ¡20 rep users to talk in a room;
don't ...	Don't tell me to review a declined flag on a deleted post; Don't allow suggested edits to be "finished" while someone has clicked "improve";

4 Experiment Setting

4.1 Dataset

To support our experiment, we dumped the dataset from meta.stackoverflow.com which is a user feedback site for Stack Overflow. We imported the XML format dataset to the database, and we analyze it and get some valuable things. The detail is showed in Table 3. As the table depicts, in the dataset, there are 5770 questions. In the whole questions, there are 909 questions being tagged "feature-request" and 4861 questions being without "feature-request" tag. The time period of the dataset is from 2008.11.26 to 2014.9.14. But we are surprised that the most of questions are posted in 2014, so we boldly believe that people are involving in the social Q&A sites quickly now.

Even though there are 909 questions being tagged "feature-request" in all 5770 questions, we still want to explore how many questions are about feature requests in the whole questions without "feature-request" tag. We randomly choose 500 questions from the untagged questions and tag the questions manually. To eliminate the divergence of personality and make the tag process incorrect, we apply a method that two authors tag the questions separately and then

Table 3. Statistics of Experiment Dataset

#questions	# "feature-request"	#no "feature-request"	time period
5770	909	4861	2008.11.26-2014.9.14

filter the questions that were tagged the same tags by both authors. Because both authors are not professional developers for Stack overflow, it is not absolutely correct in tagging questions, but the obvious questions can be tagged correctly.

In the process of tagging, we find that many questions are about features, bugs and even complaints. The result of tagging is showed in Table 4, and we can find that many questions without "feature-request" tags are also about feature requests. In all 500 questions, author 1 and author 5 labeled 181 and 121 questions about feature requests separately, the other 319 and 379 questions are not about feature requests. We integrate the result of two authors do and get 106 questions which both regard as feature requests and 304 questions which are not regarded as questions about feature requests by both authors. We spent much time doing the process of label questions manually, and it is absolutely hard to find the feature requests from the useless questions. So it is significant to propose an approach to tag the questions automatically, and our work is valuable and will assistant the developers to find the feature requests from large numbers of questions in social Q&A sites.

Table 4. Result of Tagging Questions Manually

	#questions	#about feature	#about non-feature
author1	500	181	319
author5	500	121	379
author1&author5	410	106	304

4.2 Performance Metrics

In this section, we focus on the performance metrics that will be used in our evaluation. We consider precision, recall and F-measure to evaluate the result of experiments. Precision is the ratio between returned results that are correct and the total number of returned results. Recall is the ratio between the returned results that are actual feature requests and the total number of feature requests in the input. F-measure considers both the precision and the recall, and can be interpreted as a weighted average of the precision and recall. F-measure is defined as follows in detail:

$$F = 2 \times \frac{precision \times recall}{precision + recall} \tag{4}$$

The precision, recall and F-measure can depict the performance of one approach for requirement acquisition well. For precision, recall and F-measure, the values of them are the bigger the better.

5   Results and Discussion

In this section, we introduce the experiment results in detail, comparing the performance of the approach based on linguistic rules, the approach based on pure SVM and the approach based on SVM with requirement dictionary. Then we consider the scale of testing data and test the performance in different approaches with the increasing numbers of testing data.

5.1   Comparison Among Three Approaches

In this experiment, we also choose the same 200 questions tagged "feature-request" by users and 200 questions which are labeled as "non-feature" by both two authors manually, and the whole 400 questions will be used as training examples to train the SVM model. The only difference of the two SVM experiments is whether considering the requirement dictionary about requirement acquisition. In SVM with requirement dictionary, we use the dictionary to correct the TF-IDF values of each word in documents. It is different from the previous approach based on linguistic rules, because we not only consider the titles but also the bodies of questions to train the SVM model. Our work is focusing on finding the questions about feature requests automatically and accurately.

Table 5. Result of the Experiments

	linguistic rules	pure SVM	SVM with requirement dictionary
Precision	61%	68.8%	72%
Recall	55%	75%	77%
F-measure	57.8%	71.77%	74.42%

The results of experiment is showed in Table 5 in detail. Focusing on 200 questions as testing examples, which contain 100 questions about feature requests and 100 questions are not feature requests. The results of different approaches are different, and the precision of the linguistic approach is 61%, but it can achieve 68.8% and 72% for pure SVM and SVM with requirement dictionary. The recalls among them are 55%, 75% and 77%. The F-measures for three different approaches are 57.8%, 71.77% and 74.42%. From the results we can see that the SVM approach can achieve better performance than the approach based on linguistic rules, and the improved SVM with requirement dictionary can also improve the performance of feature acquisition. The results fit our expectation. The approach based on linguistic rules is easy to understand. But depended on limited training data, it is hard to define enough linguistic rules that can fit the texts posted by users. Because users express their requirements in different types, and the process of defining linguistic rules manually is hard and time consuming. As Table 5 shows, the approach based on SVM is effective, SVM can depend on the words of every text and automatically construct a classifier based on the limited training data, but we can find that there is not much difference in the performance between pure SVM and SVM with requirement dictionary. The result is because the dictionary we build has only 74 words, it is obviously not enough to sum up the words usually used to express requirements.

5.2   The Result in Different Scales of Testing Data

Using the three different approaches in requirement acquisition, we want to know the performance with the increasing scales of testing data. Here we only consider the recall of different approaches because of the hard work in tagging questions, it also means that how many questions about feature requests can be extracted through different approaches. We separately choose 200, 300, 400, 500, 600 and 700 questions which are tagged "feature-request" by users of meta.stackoverflow. com, and test the data through 32 linguistic rules, pure SVM model and SVM model with requirement dictionary. The results is showed in Fig. 2.

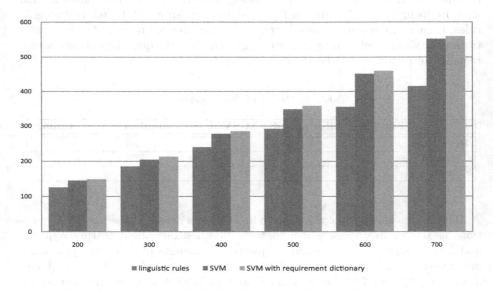

Fig. 2. Result in different scales of testing data

As Fig. 2 shows, with the increasing number of testing data, the number of questions which are extracted by three different approaches are also increasing. And we can find that the performance among three approaches remains the same trend. The performance of the approach based on linguistic rules decreases with the increasing numbers of testing data obviously, because the small number of linguistic rules can't match the more and more users' expressions.

6   Threats to Validity

The experiment results show that SVM with requirement dictionary can achieve a better result than approach based on linguistic rules and pure SVM, but there are still some threats. Firstly, the authors of this paper are not professional software developers, the manual tagged questions may be not always true for real developers. Secondly, in our experiment we choose the questions tagged "feature-request" by users as the training examples about feature requests, and this is based on the hypothesis that the users post their feature requests accurately. At last, in our experiment we increase the weight of words in documents through three times its number of occurrences, but this strategy is not perfect. So there is still a lot of work needs to be done for a better approach to mine the feature requests for assisting the software developers.

7   Conclusion

User feedback is important for software development. With the development of Internet communities, users can express their opinions through many ways, such as forums, mail-list, and software marketplace. These years have focused on the development of Stack Overflow that is a famous programming Q&A sites, and the social Q&A has become more and more popular because of its better environment. Based on the new way to gather users' feedbacks, more and more Q&A sites have been established, there are also some Q&A sites for software. But the social Q&A sites are open and free, a lot of information is invaluable, so to find a method to extract the valuable information is important. In this paper, we propose an approach based on SVM with requirement dictionary about feature acquisition. Comparing the new approach with pure SVM and the approach based on linguistic rules, we can find the approach based on SVM with requirement dictionary about feature acquisition is better than pure SVM and linguistic method. In the future, we hope to find the latent topics about requirements from users. We hope to establish a better approach to extract the most important information for continuous software development.

Acknowledgments. In this paper, the research was sponsored by the National Natural Science Foundation of China (Grant No.61432020 and No.61472430).

References

1. Vasilescu, B., Serebrenik, A., Devanbu, P., et al.: How social Q&A sites are changing knowledge sharing in open source software communities. In: Proceedings of the 17th ACM Conference on Computer Supported Cooperative Work & Social Computing, pp. 342–354. ACM (2014)
2. http://stackexchange.com/tour
3. Iacob, C., Harrison, R.: Retrieving and analyzing mobile apps feature requests from online reviews. In: Proceedings of the 10th Working Conference on Mining Software Repositories, pp. 41–44 (2013)

4. Harman, M., Jia, Y., Zhang, Y.: App store mining and analysis: Msr for app stores. In: Proceedings of the 9th Working Conference on Mining Software Repositories, pp. 108–111 (2012)
5. Galvis Carreño, L.V., Winbladh, K.: Analysis of user comments: an approach for software requirements evolution. In: Proceedings of the 35th International Conference on Software Engineering, pp. 582–591 (2013)
6. Chen, N., Lin, J., Hoi, S. C., Xiao, X., Zhang, B.: AR-miner: mining informative reviews for developers from mobile app marketplace. In: Proceedings of the 36th International Conference on Software Engineering, Hyderabad, India, 31 May–07June 2014
7. https://en.wikipedia.org/wiki/Supervised_learning
8. LIBSVM: a library for support vector machines
9. http://www.csie.ntu.edu.tw/cjlin/libsvm/
10. Noll, J.: Requirements acquisition in open source development: Firefox 2.0. Open Source Development, Communities and Quality, pp. 69–79. Springer, US (2008)
11. Bajaj K, Pattabiraman K, Mesbah A.: Mining questions asked by web developers. In: Proceedings of the 11th Working Conference on Mining Software Repositories. ACM, pp. 112–121 (2014)
12. Somprasertsri, G., Lalitrojwong, P.: Mining Feature-Opinion in Online Customer Reviews for Opinion Summarization. J. UCS **16**(6), 938–955 (2010)
13. Cleland-Huang, J., Dumitru, H., Duan, C., et al.: Automated support for managing feature requests in open forums. Communications of the ACM **52**(10), 68–74 (2009)
14. Hu, M., Liu, B.: Mining and summarizing customer reviews. In: Proceedings of the Tenth ACM SIGKDD International Conference on Knowledge Discovery and Data Mining, pp. 168–177. ACM (2004)

A Scenario Model Aggregation Approach for Mobile App Requirements Evolution Based on User Comments

Dong Sun and Rong Peng[✉]

State Key Laboratory of Software Engineering,
Computer School, Wuhan University, Wuhan, China
{rongpeng,sundong}@whu.edu.cn

Abstract. With the increasingly intense competition in mobile applications, more and more attention has been paid to online comments. For the masses, comments have been viewed as reliable references to guide the choice of applications; for providers, they have been regarded as an important channel to learn expectations, demands and complaints of users. Therefore, comments analysis has become a hot topic in both requirements engineering and mobile application development. But analyzers in both areas are always not only suffered from the vast noise in comments, but also troubled by their incompleteness and inaccuracy. Therefore, how to obtain more convincing enlightenments from comments and how to reduce the manpower needed become the research focuses. This paper aims to propose a Scenario Model Aggregation Approach (SMAA) for analyzing and modeling user comments of mobile applications. By selecting appropriate natural language processing technologies and machine learning algorithms, SMAA can help requirements analysts to build aggregated scenario models, which can be used as the source of evolutionary requirements for the decision making of application evolution. The aggregated scenario model is not only easy to read and understand, but also able to reduce the manpower needed greatly. Finally, the feasibility of SMAA is exemplified by a case study.

Keywords: Scenario model aggregation approach · Aggregated scenario model · Mobile application · User comments · Kernel concerns

1   Introduction

With the increasing development of mobile Internet, mobile applications become more and more prevalent. But intense competition, short life cycle and low user adhesion are all obstacles for their success. Being alert to the changes of user expectations and evolving the mobile app accordingly are the only way for them to stand out. Therefore, online comments, as one of the most important channels of expressing user expectations and dissatisfaction, have become a vital source to obtain evolutional requirements from the masses.

The contents described in user comments always have strong relationships with scenarios. But as online comments are spontaneously described by users rather than elaborately elicited by professional requirements engineer, most of them do not contain the essential scenario information clearly, which makes them difficult to

L. Liu and M. Aoyama (Eds.): APRES 2015, CCIS 558, pp. 75–91, 2015.
DOI: 10.1007/978-3-662-48634-4_6

understand and results in incomplete and inaccurate understanding on the real intents of users. To help requirements analysts understand authentic user intents, it needs to develop a scenario extraction, modeling and aggregation method to extract and aggregate scenario information hidden inside those comments.

This paper proposes a Scenario Model Aggregation Approach (SMAA) which can support different components to utilize various natural language processing (NLP) technologies, machine learning algorithms and modeling methods to analyze and categorize comments and their attached records, and construct aggregated scenario models (ASMs) based on certain kernel concerns. Each part of SMAA is exemplified by a sample method. Finally, the feasibility of SMAA is shown by a case study.

The remainder of this paper is organized as follows: Section 2 introduces related work. Section 3 introduces the preliminary knowledge. Section 4 elaborates SMAA with sample methods. Section 5 presents the case study. Section 6 presents the comparison with related work.   Section 5 draws out the conclusion and further work.

2   Related Work

As the development of mobile application stores, many studies focus on the analysis and utilization of online comments of mobile applications. These studies can be mainly divided into two categories.

One category is concerned on the types [1, 2], characteristics [1] and effects of App comments [3-5]. They point out that some types of comments, such as functional errors and requests for additional features, are good sources to obtain user requirements [1, 2]. But how to extract requirements is not the focus of these articles.

The other category is concerned on how to extract and analyze information from App comments. We summarized the research topics, research methods, and outputs of this category in Fig. 1.

The 1st research topic is **informative comments extraction**. As users post their comments for different purposes, some comments are noises from the perspective of requirements extraction. Therefore, many researchers [6, 7] have focused on how to extract informative comments from the raw comments. Thus, the input of this stage is raw user comments, and the output is informative comments in the perspective of requirements engineers. The informative comments extraction mainly adopts classification models or keyword based extraction methods. E.g., Bayesian classifier is used to extract informative user comments by filtering noisy and irrelevant ones [6]. A prototype system is developed to extract new feature requests by summarizing 237 keyword based grammar rules [7].

The 2nd research topic is **requirements topics extraction**. Extracting requirements topics from thousands to millions of comments manually is time-consuming and laborious. So, many researchers are dedicated on how to extract feedback topics from user comments automatically. The input of this stage is informative comments, and the outputs are representative sentences or words of each requirements topic. Topic models and cluster models are commonly used in this stage. E.g., topic models such as LDA [6-9] and ASUM [10] have been used to extract implicit topics and representative topic words from informative comments. Clustering models, such as K-Means

[11] and GN community discovery model [12], have been used to cluster comments and select the topic sentence for each cluster.

The 3rd research topic is **Topic-based analysis**. After requirements topics extraction, developers try to use the topics for further analysis. The input of this stage is usually requirements topics and their related data, the outputs are various analysis results for decision-making. Statistical analysis methods are commonly used in this stage. For example, granger causality model are used to analyze whether the utility of a topic (system aspect) is useful in forecasting the software sales [11]. Linear regression model are used to identify reasons why users like or dislike a given App [8].

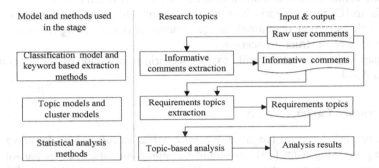

Fig. 1. Research topics of information extraction and analysis from App comments

The above studies focus on extracting useful information and topics from online user comments. For the providers, the extracted information can be used to learn users' expectations and complaints.   But as pointed in [13], due to the arbitrariness and imperfection of the expression, it would be difficult to spy out the true intent behind the text. Thus, only relying on the surface meanings of the online comments to infer user requirements and make evolution decisions will face significant risks.

This paper focuses on providing a Scenario Model Aggregation Approach (SMAA) which can integrate the above techniques to construct ASMs from user comments and related data sources, which can help analyzers learn the implications more clearly and accurately.

3   Preliminaries

3.1   Definitions

Definition 1: Raw User Comment (RUC)
RUCs refer to the raw comments which are posted by users of some specific mobile application in a certain application market. They usually contain the information such as text, rating, and publishing time. Some comments may present user preferences, error feedbacks and advices which are useful to understand users' demands. And others may only contain useless information from the perspective of requirements engineers such as pure emotion expressions and advertisements.

Definition 2: Informative Comment (IC)

ICs refers to the comments which contain useful information for further improvement of the application, such as new feature requests, defect feedbacks and error reports. ICs are also called as potential evolution requirements, as they can be used to extract user requirements. According to the different intents, ICs can be divided into 2 categories: Improvement Comments (ImCs) and Fault Feedback Comments (FFCs).

Definition 3: Kernel Concern (KC)

KC refers to the core user demand implicated in a specific IC.

Definition 4: Informative Comment with Scenario Information (SIC)

SICs refer to ICs with scenario information. Scenario information refers to the information, such as trigger operations, usage contexts and underlying rationales, which can be used to reconstruct the scenario.

This paper aims to automatically extract scenario information of similar comments and build ASMs.

3.2 Stanford Parser

Stanford parser is a multi-language syntactic parser developed by Stanford natural language processing team [14]. Its output has many formats, such as part-of-speech (POS) tagged text, phrase structure tree (PST), and dependency relations (DRs) [14]. POS, PSTs and DRs are always adopted by various methods to understand the text written in natural language automatically.

In this paper, firstly, we use Stanford Word Segmenter to split each comment written in Chinese into a sequence of words; then, the sequence is imported into Stanford parser to get its PST and DRs for further processing.

3.3 Decision Tree Model

Among various classification models, decision tree model (DTM) is famous for its good understandability and high accuracy. DTM is a tree-like model used to predict or classify. Each non-leaf node represents a "test" on an attribute and each leaf node represents a class label. Each branch represents the outcome of the test and the path from root to a specific leave represents classification rules.

In SMAA, it recommends using DTM to identify SICs. The dependency relations identified by Stanford parser and the category attribute identified by analysts will be used to construct DTM.

4 Scenario Model Aggregation Approach and Its instantiation

4.1 Scenario Model Aggregation Approach

To help requirements engineers to extract the scenario information and aggregate ASMs from RUCs, a kernel concern based scenario model aggregation approach is proposed. The approach includes the following 5 stages, as shown in Fig. 2:

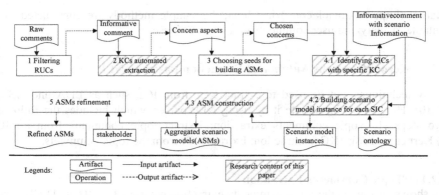

Fig. 2. The main steps of SMAA

1) Filtering RUCs: In this stage, the goal is to obtain ICs from RUCs through de-noising and filtering. Many machine learning algorithms can be used to fulfill the task, such as Naive Bayes [6, 15].

2) KCs automated extraction: This stage aims to mine representative KCs from the collection of ICs. NLP and data mining technologies can be used in this stage, which will be exemplified in Section 4.2.

3) Choosing seeds for building ASMs: In this stage, an analyst needs to choose KCs they are interested in, and take these concerns as seeds to construct ASMs.

4) Constructing ASMs for each KC: In this stage, the following operations should be performed on each chosen KC:

    4.1) Identifying SICs with a specific KC: Find all ICs which contain the KC and identify those with scenarios information to construct a set of SICs. Various classification approaches, such as decision tree and deep learning, can be used. The sample method is described in Section 4.3.1.

    4.2) Building scenario model instance for each SIC: For each SIC, extract the scenario information from the text of the SIC and its attached record, such as device type and OS type; and then create a scenario model instance. Ontology-based and context-aware methods can be used to extract scenario information automatically. An ontology-based scenario model instance construction method is exemplified in Section 4.3.2.

    4.3) ASM construction: Aggregate all the scenario elements from the scenario model instances of a specific KC and build a ASM. The way of aggregating scenario model instances should be decided according to the characteristics of the scenario elements. For instance, "AND/OR tree" can be used for aggregating scenario elements of "trigger condition".

5) ASMs refinement: The kernel concern based ASMs is checked one by one manually to verify whether it misses some key elements or not. If the information of any necessary element is deficient, the analyst should organize the relative stakeholders together to refine the models. The refined model can be used as the reference model, which can help providers and analyzers to understand the user demand correctly.

Due to space limitation, the steps 1, 3 and 5 will not be discussed further as many existing methods can be used directly. In the following sections, only the sample methods of step 3 and 4 will be elaborated. Since the sample data are from Android

Market (http://apk.hiapk.com/) in China, the sample methods are constructed to be suitable to analyzing user comments written in Chinese.

4.2   Kernel Concerns Automated Extraction Based on NLP

Many techniques can be used to extract KCs from ICs, such as LDA and ASUM [6-10]. Here, we'll instantiate it by using NLP parser, which includes the following two steps: 1) Topic comments extraction: Extract topic comments from all ICs; 2) Kernel concerns (KCs) extraction: Extract KCs from the topic comments.

4.2.1   Topic Comments Extraction

As shown in Fig. 3, the topic comments extraction contains two steps: 1) Clustering ICs; and 2) choosing representative topic comments (RTCs) for each cluster.

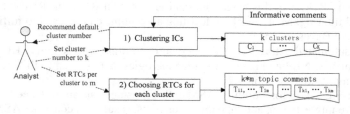

Fig. 3. Topic comments extraction

1) Clustering ICs: Using K-means can classify ICs to several clusters. The recommended cluster number (RCN) can be calculated according to a modified version of Cans' metric [17, 18], in which RCN will be high if individual comments are dissimilar with each other, and low otherwise. If all comments are same, RCN equals to 1. RCN should be equal or lesser than the number of the comments. Analysts can manually adjust RCN according to their own will or reading ability: the bigger the cluster number is, the more the representative topic comments they need to read [18].

2) Choosing RTCs for each cluster: After the analyst determines the number of RTCs for each cluster (here suppose it to m), the comments which is m nearest to the centroid (measured by cosine similarity) will be chosen as the representatives of the cluster.

4.2.2   Kernel Concerns Extraction

Here, feature requests and fault feedbacks are regarded as the default KCs recommended automatically from the perspective of requirements engineers. And analysts can manually modify the KCs according to their preferences.

By analyzing the features of output formats of Stanford parser, the KCs extraction algorithm can be designed as follow:

Algorithm 1: KCs extraction algorithm (KCsEA)
Input: representativeTopicComment
Output: kernelConcern
1) Parse **representativeTopicComment** by Stanford Parser to get its phrase struc-

ture tree **GT** and dependency relations set **DependencySet**;

   2) Identify the core word of verb phrase **VP**:

      2.1) Locate in the bottom right **VP** in the **GT**;

      2.2) Annotate the last verb (**VV** or **VA**) of this **VP** as its core word;

   3) Expand the core word: Find the words in **DependencySet** which has one of the following dependency relations with the core word: **advmod** (adverbial modifier), **nsubj** (nominal subject), and **dobj** (direct object); and combine the located modifier with the core word to construct the output **kernelConcern**;

   4) return **kernelConcern**.

The returned **kernelConcern** will be regarded as recommended KC for the specific topic comment. It will be displayed together with the topic comment to analysts, which can help them understand its meaning and context. E.g., the KC of RTC "升级后经常死机" (in English "often crashes after upgrade") is shown as Fig. 4.

In Chinese: | "升级后经常死机" |
In English: | "**often crash** after upgrade" |

Fig. 4. The display style of a RTC and its recommended KC

Recommending KCs automatically can greatly reduce the manual workload of extracting them from a large volume of ICs. At the same time, it allows the analyzer to modify or redesignate the KCs according to his/her preference, which is helpful to construct ideal scenario models.

4.3   KC Based ASM Construction

Many modeling methods can be used to construct ASMs based on KCs. The following method is just a sample method.

4.3.1   Decision Tree Based Automatic Identification of Informative Comments with Scenario Information

As stated in Section 3.3, the paths from root to leaf represent classification rules in decision tree model. These paths are helpful to understand which dependency relation has closely correlation with scenario information. Therefore, decision tree model is chosen as the classification model to distinguish ICs with or without scenario information.

   As shown in Fig. 5, the identification process of SICs based on decision tree model has two stages: constructing classifier and using classifier.

   The process of **constructing classifier** is as follow:

   1) Annotating the categories: Select some comments as the training dataset, and ask some analyst(s) to pick out all SICs. Thus, all training data will have a class label. Table 1 showes some sample ICs annotated by an analyst.

   2) Parsing IC: Get the dependency relations of each IC in the training dataset by Stanford parser.

   3) Vectorizing IC: Use dependency relations and the category information to

construct attribute vectors for ICs: $ca = (d_1, d_2, ..., d_n, r)$, where d_i ($i=1..n$) represents a certain dependency relation, n is the number of frequently used dependency relation in Chinese [14], and r is its corresponding class label.

For example, suppose that the kth IC in the training dataset contains dependency relations 1, 2 and 3, and it is annotated as SIC. Then its attribute vector is: $ca_k = (1,1,1, 0, ...,0, 1)$

4) Training the model: Select the C4.5 algorithm in WEKA 3.12 to train the model.

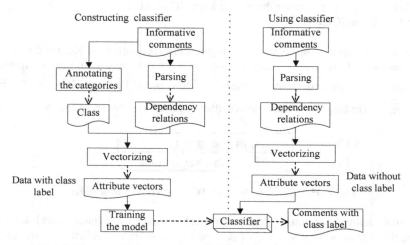

Fig. 5. The identification process of SICs based on decision tree model

The process of **using classifier** is as follows:

First parse all ICs one by one to get the dependency relations of each comment; then, construct the attribute vector $cd_i = (d_{i1}, d_{i2}, ..., d_{in})$ for each IC; finally, use the trained model to test all vectors to determine whether they are SICs or not.

Table 1. ICs with or without annotation

NO	Comments in Chinese (C) and English (E)	SIC?
1	打开定位后就自动退出 (C) Abnormal exit after open the positioning function (E)	☑
2	有时候自动退出，望改进 (C) Sometimes it will exit abnormally, please fix it.(E)	☐
3	老是闪退 (C) Always quit without prompt.(E)	☐
4	5.0版本有时出现闪退 (C) Version 5.0 always exits without prompt.(E)	☑

4.3.2   Scenario Model Instances Construction

As pointed in [16], scenario model can help requirements analysts understand the authentic intentions of users. Thus, building scenario model is crucial for extracting scenario information from comments and their attached records.

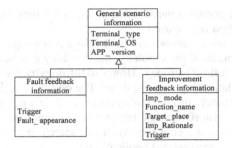

Fig. 6. Scenario metamodel

As shown in Fig. 6, a scenario metamodel is defined according to the available scenario information which could be extracted from the features of user comments in mobile application stores. The scenario metamodel defines that a scenario model should contain the following information:

1) General scenario information: General scenario information contains some basic elements such as terminal type, terminal OS and application version. These kinds of information are recorded by mobile applications as soon as users post their comments. Thus, they can be crawled or directly accessed from the application stores.

2) Categorized information: the information extracted from the text of the comment. As Fault Feedback Comments (FFCs) and Improvement Comments (ImCs) are 2 typical kinds of comments, the information extracted from them should be also classified into 2 categories: improvement feedback (IF) info and fault feedback (FF) info.

The scenario information in FFCs mainly concerns on: the trigger (the operation(s) triggered the fault) and the fault appearance (system appearance when fault occurs).

The scenario information in ImCs mainly focuses on: the improvement mode (added/ modified / deleted / improved / lowered, et al), the specific function name which need to improve, the target place (usually represented by its parent function), the improvement rationale and the trigger. The data models of 2 kinds of scenario model instances are shown in Fig. 7(a) and 7(b).

```
FFScenarioModel {
// General Info
Terminal terminal;
OS os;
AppVersion appVersion;
// FaultFeedback ScenarioInfo
Type type=faultFeedback
KernelConcern kernelConcern;
Trigger trigger;
FaultAppearence faultAppearence;
}
```

(a) Data model of FF Scenario Model

```
IFScenarioModel {
// General Info
Terminal terminal;
OS os;
AppVersion appVersion;
// ImprovementFeedback ScenarioInfo
Type type=improvementFeedback
KernelConcern kernelConcern;
Trigger trigger;
ImpMode impMode;
FuncName funcName;
Target target;
ImpRationale impRationale;
}
```

(b) Data model of IF Scenario Model

Fig. 7. Data models of scenario model

Based on the above scenario metamodel, an ontology based scenario model information extraction method is proposed.

The ontology tree model is shown as Fig. 8. The top ontology is the root. It

According to the source of the ontology, the ontology can be divided into **DomainOntology** and **ApplicationOntology**. **DomainOntology** contains the domain general concepts and the relationships among them. The concepts such as "add", "modify" and "delete" are all instances of DomainOntology, as they are general concepts for the whole domain. **ApplicationOntology** represents the ontology related to a specific application. For example, the concepts, such as "search around" and "positioning", may only be used for the map related applications.

According to the nature of the ontology, it can be divided into **ActionOntology**, **EntityOntology** and **ModifierOntology**. **ActionOntology** can be further subdivided into **OperationOntology**, such as "click" and "move", and **ExpectationOntology**, such as "hope" and "suggestion". **EntityOntology** can be subdivided into **FunctionOntology**, such as "navigation" and "positioning", **PeripheralOntology**, such as "camera" and "microphone", and **FaultOntology**, such as "exception" and "crash".

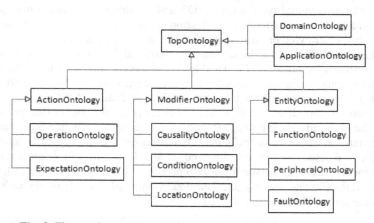

Fig. 8. The ontology tree model for scenario model element extraction

The scenario model instance construction algorithm based on ontology tree is as follow.

Algorithm2: Scenario model instance construction algorithm (SMICA)

Input: *ontoModel*; // the ontology tree model

    sic; //the comment with scenario information

Output: *sic.scenarioModel* //the scenario model instance of *sic*.

1) Parse *sic* by Stanford parser to get the phrase structure tree *sic.struTree,* and the dependency relations set *sic.DependencySet.*

2) Scan the leaf nodes in *sic.struTree* one by one according to the *ontoModel*: if a leaf node matches one of the concepts in the **ontoModel**, the node will be marked with the ontology category label; otherwise it is marked with label "**Others**".

3) Construct scenario model instance according to *sic*:

    3.1) When *sic* contains a phrase marked with ExpectationOntology, extract scenario elements according to the data model of IF scenario model, and construct the corresponding model instance *sic.ScenarioModel*:

    3.1.1)   *sic.scenarioModel.type= improvementFeedback* ;

    3.1.2) Find the leaf node which is marked with ExpectionOntology, and up traverse *sic.struTree* from this node to find the nearest *vp* which contain a phrase marked with OperationOntology, and assign the verb (*v*) to the element **impMode** : *sic.scenarioModel.impMode* = *v*, and assign the verb's nearest direct object *e* to the element **funcName**: *sic.scenarioModel.funcName* =*e*;

    3.1.3) If there exists a leaf node *fn* in *sic.struTree* which is marked with FunctionOntology, check whether its parent node has a leaf node labeled as LocationOntology. If yes, assign *fn* to element "**Target**": *sic.scenarioModel.target=fn*;

    3.1.4) If there exists a leaf node which is marked with ConditionOntology, find the word *tp* which has a dependency relation **case**, and assign *tp* to the element "**trigger**": *sic.scenarioModel.trigger=tp* ;

    3.1.5) If there exists a leaf node which is marked with CausalityOntology, find the word *tp* which has a dependency relation **case**, and assign *tp* to the element "**impRationale**": *sic.scenarioModel.impRationale* = *tp*;

    3.2) When *sic* contains a phrase marked with FaultOntology, extract scenario elements according to the FF scenario metamodel, and construct its model instance *sic.scenarioModel* accordingly as follows:

    3.2.1)   *sic.scenarioModel.type= faultFeedback;*

    3.2.2) Locate the leaf node *fo* which is marked with FaultOntology; expand the word *fo* with its nearest adjunct word and record the expand phrase as *efo*, and assign it to the element "**faultAppearence**": *sic.scenarioModel. faultAppearence* = *efo*;

    3.2.3) If there exists a leaf node which is marked with ConditionOntology, find the word *tp* which has a dependency relation **case**, and assign it to the element "**trigger**": *sic.scenarioModel.trigger* = *tp* ;

    3.3) Automatically extract the general scenario elements of *sic* and assign them to the corresponding model elements;

4)   Return *sic.scenarioModel*.

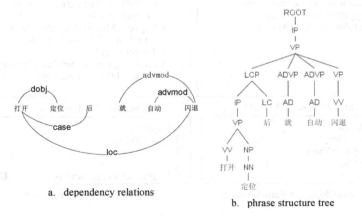

a. dependency relations

b. phrase structure tree

Fig. 9. A parsing example

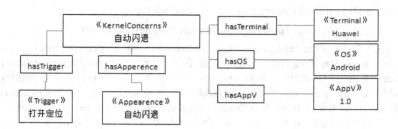

Fig. 10. The corresponding scenario model instance of Fig. 9

For example, the comment "打开定位后就自动闪退" ("Quit without prompt after open the positioning function" in English) can be parsed by Stanford parser to obtain the dependency relations and the phrase structure tree as shown in Fig. 9(a) and 9(b). Referred to the ***ontoModel***, the leaf nodes "打开"(Open)、"后"(After)、 "闪退"(Quit without prompt) are marked with OperationOntology, LocationOntology, and FaultOntology respectively. Therefore, its model instance can be constructed according to the above algorithm shown in Fig.10.

4.3.3　Aggregated Scenario Model Construction

Aggregated scenario models are also classified into 2 categories: the improvement feedback aggregated scenario models (IFASM) and the fault feedback aggregated scenario models (FFASM), as shown in Fig. 11(a) and 11(b). Where,

ASM {
// General Info
CTerminalSet terminalSet;
COSSet osSet;
CAppVersionSet appVersionSet;
// fault feedback aggregation
Type type=faultFeedback
CKernelConcern kernelConcern;
CTriggerSet triggerSet;
CFaultAppSet faultAppSet;
}
	ASM {
	// General Info
	CTerminalSet terminalSet;
	COSSet osSet;
	CAppVersionSet appVersionSet;
	// improvement feedback aggregation
	Type type=improvementFeedback
	CKernelConcern kernelConcern;
	CTriggerSet triggerSet;
	CImproveModeSet improveModeSet;
	CFuncNameSet funcNameSet
	CTarget target
	CFaultAppearenceSet faultAppearenceSet;
	CRationaleSet rationaleSet
	}

(a) Data model of FFASM	(b) Data model of IFASM

CKernelConcern {
Dimension dim;
DimValue dimValue
Times times;
}
	CTrigger {
	Trigger trigger;
	Times times
	}
	CTriggerSet triggerSet= new Set of CTrigger

(c) Data Model of CKernelConcern	(d) Data Model of CTriggerSet

Fig. 11. Data models of ASMs

1) Type: Represent the type of the aggregated scenario model;

2) CKernelConcern: As shown in Fig. 11(c), it has 3 attributes: Dimension, Dim-Value and Times. Dimension represents the designated aggregate dimension which can be any dimension in the scenario model instance; DimValue is a specific value the aggregation focuses on; Times is a counter which record the number of SICs aggregated under the condition of the same Dimension with same DimValue.

3) CTriggerSet: Trigger is an attribute in scenario model instance. It records the trigger of the comment. Accordingly, CTriggerSet records the triggers of all aggregated model instance. For each element CTrigger in CTriggerSet, it consists of the attribute trigger and times, where trigger is the same as the one in the model instance, and times is the count of model instances with a certain same trigger.

4) Other attributes: The data structure of each set attribute is similar with CTriggerSet.

By using the above data structure, the following algorithm can be used to aggregate the model after all model instances are available.

Algorithm 3: ASM construction algorithm

Input: insScenarioModelSet; // the scenario model instance set to be aggregated

insModelType; // the type of model instances to be aggregated

dim, dimvalue; // the dimension and its value to be aggregated

Output: m; // the aggregated scenario model

1) New an empty ASM:

m = new ASM (insModelType);

2) Set the attributes of the Kernel Concern:

m.kernelConcern.dim=dim;

m.kernelConcern.dimValue=dimValue;

m.kernelConcern.times=0;

3) Traverse the **insScenarioModelSet**:

For each model instance *c* in **insScenarioModelSet**, execute:

if(*c.scenarioModel.dim*==*dimvalue*&&*c.scenarioModel.type*==*m.type*) then

{   m.kernelConcern.times++;

for any *d* different from Dimension{*dim*, *type*} in **c**

{ if exist (*c.scenarioModel.d*, *m.d*) then

{*e*=getelem(*m.d*, *c.scenarioModel.d*); *e.times*++; putelem(*m.d*, e)}

else { *e*=newelem(*c.scenarioModel.d*, 1); add(*m.d*, e)}

4) return **m**.

Through the above sample methods, the SMAA can be implemented to support the requirements engineers to analyze the kernel concerns hidden in raw user comments and construct aggregated scenario models.

5   Case Study

A mobile application named "Baidu Map" in Android Market (http://apk.hiapk.com/) is selected to exemplify the effect of SMAA. Baidu Map has 2335 user comments in this store by January 1, 2014. Following the steps in SMAA strictly, the results is as follows:

1) Filtering RUCs: Using Naive Bayes and down sampling method described in [15], 610 ICs are obtained.

2) KCs automated extraction:

2.1) Topic comments extraction: According to section 4.2.1, the K-means clustering algorithm in Weka 3.6.10 is adopted to cluster the ICs. The cluster number is set to 11 based on Can's metric [18], and the distance function adopted is Euclidean distance. The nearest comment to the centroid is selected as the RTC for each cluster, which is listed in the 2nd column of Table 2.

2.2)Kernel concerns extraction: According to section 4.2.2, the KCs extraction algorithm based on Stanford parser is adopted to extract and recommend kernel concerns for each topic comment. These concerns are listed in the 3rd column of Table 2 and the final accepted KCs by the authors are listed in the 4th column.

3) Choosing seeds for building ASMs: Choose "闪退" ("quit without prompt" in English) as the seed, namely the selected KC to construct and refine the ASM.

4) ASM construction based on the KC:

4.1) Identifying SICs with the KC: Find all ICs which contain the chosen seed "闪退(quit without prompt)" through retrieval. 15 ICs are retrieved, and 11 of them are labeled as SICs. The decision tree model is trained by C4.5 algorithm in Weka 3.6.10.

4.2) Building scenario model instance for each SIC: According to the algorithm proposed in section 4.3.2, the scenario model instances are built for each comment;

4.3) ASM construction: According to the algorithm in section 4.3.3, the aggregated scenario model with the kernel concern "闪退(quit without prompt)" is built as shown in Fig. 12.

Table 2. Comments Analysis for Baidu Map in Andriod Market

CID	Representative comments of each cluster	Recommended KCs	Accepted KCs
1	4.0不能用 4.0 cannot be used	不能用 Cannot be used	
2	百度地图这弹广告不说,而且不能退出老是自动启动 Not to mention pop ads, Baidu map exit abnormally and often restart.	退出老是自动启动 exit abnormally and often restart	自动启动 Restart
3	新版本4.5，闪退 New version 4.5, quit without prompt	版本闪退 Version quit without prompt	闪退 Quit without prompt
4	度娘又升级了，误差增加到十千米。 Baidu update again, error increased to 10 kms	增加误差 error increase	误差增加 Error increased
5	我是note2，装好后不运行，但电量里看百度地图耗电最高 Note2, not running after installed, but its power consumption is the highest	耗电最高 Highest power consumption	耗电 Power consumption
6	地图不准 Map is not accurate	地图不准确 Map is not accurate	
7	耗电量大，手机发热厉害，都快50℃了 Large power consumption, it become very hot, almost 50℃	50℃了 Almost 50℃	发热 Become hot

Table 3. (*Continued*)

8	定位不准 Positioning is not accurate	定位不准 Positioning is not accurate	
9	地图更新太慢！ Map update too slow!	**地图更新太慢** Map update too slow	
10	定位太差了! Positioning is too bad!	定位太差 Positioning is too bad!	
11	没有农村离线地图。 No rural offline map.	离线地图 No offline map	

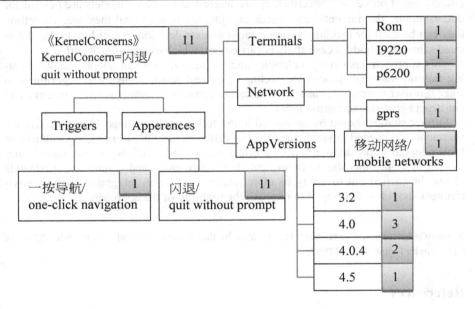

Fig. 12. The ASM in the concern of "闪退(quit without prompt)"

According to the ASM, a requirements analyzer can clearly learn that the version 4.* often "闪退" (quit without prompt), one of its trigger operations is "一按导航" (click navigation), and occurred in GPRS and mobile networks. But the type of the terminal is not clear. If needed, the analyzer can launch a refinement activity according to the step 5 in SMAA.

6   Comparison with Related Work

As summarized in section 2, previous work focused on how to use specific technology to extract and analyze information form App comments. The main outputs of previous work were specific information such as informative comments, topic words, topic sentences, and analysis results based on topic.

This focuses of the proposed approach SMAA are that 1) automated extract scenario information from comments, and 2) organize the information in appropriate forms for good visibility and understandability. The outputs of SMAA are ASMs which are suitable for evolution requirements analysis and refinement.

7   Conclusion and Future Work

This paper proposes a scenario model aggregation approach SMAA with the following functionality: 1) It can support raw user comments filtering and kernel concerns extraction; 2) it can help requirements engineers to construct aggregated scenario models based on kernel concerns they are interested in. These models are helpful for understanding the authentic user needs and judging how critical they are. Therefore, they can be used for making evolution requirements decision. According to the elaborated sample methods, a case study is carried out, which exemplifies its effects.

This approach has strong scalability and flexibility, as it can adopts various existing natural language processing technologies and machine learning algorithms to filter raw user comments, classify informative comments, extract kernel concerns, and construct aggregate scenario models.

Further studies should be conducted in the following directions: 1) quantitative assessments are needed to verify the effect of different machine learning algorithms and tools adopted in the approach; 2) the effectiveness and efficiency of the sample methods and algorithms should be evaluated; 3) more cases and empirical experiments should be carried out to verify its effectiveness; and 4) an integrated analysis environment should be developed to improve the usability of the approach.

Acknowledgements. The research is supported by the National Natural Science Foundation of China under Grant No. 61170026.

References

1. Pagano, D., Maalej, W.: User feedback in the appstore: an empirical study. In: 2013 21st IEEE International Requirements Engineering Conference (RE), pp. 125–134 (2013)
2. Khalid, H.: On identifying user complaints of iOS apps. In: 2013 35th International Conference on Software Engineering (ICSE), pp. 1474–1476 (2013)
3. Harman, M., Jia, Y., Zhang, Y.: App store mining and analysis: MSR for app stores. In: Proceedings of the 9th IEEE Working Conference on Mining Software Repositories, pp. 108–111 (2012)
4. Kim, H.-W., Lee, H.L., Son, J.E.: An exploratory study on the determinants of smartphone app purchase. In: The 11th International DSI and the 16th APDSI Joint Meeting, Taipei, Taiwan (2011)
5. Chia, P.H., Yamamoto, Y., Asokan, N.: Is this app safe? a large scale study on application permissions and risk signals. In: Proceedings of the 21st International Conference on World Wide Web, pp. 311–320 (2012)

6. Chen, N., Lin, J., Hoi, S.C.H., Xiao, X., Zhang, B.: AR-Miner: mining informative reviews for developers from mobile app marketplace. In: Proceedings of the 36th International Conference on Software Engineering, pp. 767–778 (2014)
7. Iacob, C., Harrison, R.: Retrieving and analyzing mobile apps feature requests from online reviews. In: 2013 10th IEEE Working Conference on Mining Software Repositories (MSR), pp. 41–44 (2013)
8. Fu, B., Lin, J., Li, L., Faloutsos, C., Hong, J., Sadeh, N.: Why people hate your app: making sense of user feedback in a mobile app store. In: Proceedings of the 19th ACM SIGKDD International Conference on Knowledge Discovery and Data Mining, pp. 1276–1284 (2013)
9. Oh, J., Kim, D., Lee, U., Lee, J.-G., Song, J.: Facilitating developer-user interactions with mobile app review digests. In: CHI 2013 Extended Abstracts on Human Factors in Computing Systems, pp. 1809–1814 (2013)
10. Galvis Carreño, L.V., Winbladh, K.: Analysis of user comments: an approach for software requirements evolution. In: Proceedings of the 2013 International Conference on Software Engineering, pp. 582–591 (2013)
11. Jiang, W., Ruan, H., Zhang, L.: Analysis of economic impact of online reviews: an approach for market-driven requirements evolution. In: Zowghi, D., Jin, Z. (eds.) APRES 2014. CCIS, vol. 432, pp. 45–59. Springer, Heidelberg (2014a)
12. Jiang, W., Ruan, H., Zhang, L., Lew, P., Jiang, J.: For user-driven software evolution: requirements elicitation derived from mining online reviews. In: Tseng, V.S., Ho, T.B., Zhou, Z.-H., Chen, A.L., Kao, Hung-Yu. (eds.) PAKDD 2014, Part II. LNCS, vol. 8444, pp. 584–595. Springer, Heidelberg (2014b)
13. Kompan, M., Bieliková, M.: Context-based satisfaction modelling for personalized recommendations. In: 2013 8th International Workshop on Semantic and Social Media Adaptation and Personalization (SMAP), pp. 33–38 (2013)
14. Chang, P.-C., Tseng, H., Jurafsky, D., Manning, C.D.: Discriminative reordering with Chinese grammatical relations features. In: Proceedings of the Third Workshop on Syntax and Structure in Statistical Translation, pp. 51–59 (2009)
15. Ni, Y., Peng, R., Sun, D., Lai, H.: Potential evolution requirements detect method based on user comments. Wuhan Univ. (Nat. Sci. Ed.) **61**, 347–355 (2015)
16. Sutcliffe, A.: Scenario-based requirements engineering. In: Proceedings of 11th IEEE International Requirements Engineering Conference, 2003, pp. 320–329 (2003)
17. Davril, J.-M., Delfosse, E., Hariri, N., Acher, M., Cleland-Huang, J., Heymans, P.: Feature model extraction from large collections of informal product descriptions. In: Proceedings of the 2013 9th Joint Meeting on Foundations of Software Engineering, pp. 290–300 (2013)
18. Can, F., Ozkarahan, E.A.: Concepts and effectiveness of the cover-coefficient-based clustering methodology for text databases. ACM Trans. Database Syst. **15**, 483–517 (1990)

Requirements Processes and Specifications

Extreme Requirements Engineering (XRE)

Naveed Ikram[1(✉)] and Sonia Naz[2]

[1] Riphah International University, Sector I-14, Islamabad, Pakistan
naveed.ikram@riphah.edu.pk
[2] International Islamic University, Sector H-10, Islamabad, Pakistan
sonia.msse355@iiu.edu.pk

Abstract. The importance of requirements engineering process in success or failure of software projects has tempted the organizations to improve their RE processes. In our previous study, we assessed the relative perceived values of RE practices by conducting a global survey of practitioners. The survey revealed six RE practices that were perceived as having high values by RE experts worldwide. These practices were related to stakeholder's consultation, requirements specification, and requirements management. In this paper, we are presenting an RE approach Extreme Requirements Engineering (XRE) which is based on these valuable RE practices and feedback from the practitioners on the role of customer representative. XRE complements the existing agile methods, Scrum and XP (eXtreme Programming) with six guidelines for agile teams. These guidelines ensure the extreme use of the valuable RE practices. The XRE can be helpful for practitioners to overcome many challenges faced by RE in agile.

Keywords: Index terms · requirements engineering · Agile software development · Requirements engineering practices · Agile requirements

1   Introduction

Requirements engineering plays an important role in the success or failure of software products. Most common, time-consuming and expensive to repair errors are mostly a consequence of inadequate requirements engineering process [22]. The importance of requirements engineering process has tempted the organizations to invest in the improvement of their RE processes.

The traditional RE process consists of requirements elicitation, analysis, documentation, validation, and management. Agile software development has emerged in the last decade as new and different way of software development as compared to the traditional one. The agile teams focus on iterative development and face to face communication. The product under development is divided into releases and work of every release is carried out in multiple sprints. Each sprint is typically of 2-4 weeks long and delivers an increment of functionality at the end. The requirements in the agile development emerge and evolve continuously during a project. This iterative and evolutionary nature of agile makes the RE activities challenging for agile teams. The main objective of agile is to do only most required activities that consume fewer resources and produce the maximum return. Hence, it will be appropriate if we introduce only most valuable and most required RE activities in agile.

© Springer-Verlag Berlin Heidelberg 2015
L. Liu and M. Aoyama (Eds.): APRES 2015, CCIS 558, pp. 95–108, 2015.
DOI: 10.1007/978-3-662-48634-4_7

The Scrum and XP are two most common methods of agile development. The purpose of our research is to guide Scrum and XP teams on what RE practices are suitable for them and how they can use these valuable RE practices to get high benefits and avoid RE related problems. By implementing these practices in agile, they can get maximum benefits by putting minimum effort.

The eXtreme Programming (XP) is an agile approach to software development. The philosophy of XP is to carry out good software development practices all the time and hence is called eXtreme. We adopting the same philosophy for requirements, in this paper present eXtreme Requirement Engineering (XRE) an agile RE approach. The XRE consists of practices and guidelines for agile teams. The XRE is based on most valuable practices of RE identified as a result of a global survey of RE practitioners. We have reported the details of survey and results in [1]. We conducted an online discussion with 23 agile practitioners from different countries to understand the role of product owner/customer representative. We analyzed these valuable practices in the light of discussion with agile practitioners and developed a set of practices along with guidelines to incorporate it in SCRUM framework. We, in this paper, present findings from the discussion with agile practitioners. We, also, show how valuable RE practices may be used within an agile approach in an extreme manner.

2   Background

The traditional requirements engineering requires up-front and detailed requirements thus lacks agility. A set of documents consisting of detailed requirements and models are essential in the traditional RE. Minimal requirements documentation and more face-to-face communication is one aspect of agile software development [2]. Instead of doing conventional requirements documentation agile teams do to-the-point, precise user story oriented documentation [25]. However, minimal documentation in agile approaches results into few problems such as long term maintenance of software products [3], [4]. User stories are complemented with more detailed artifacts to overcome this challenge [5] which helps in making right implementation choices at the coding stage.

Frequent collaboration with an onsite customer to avoid disagreement and difference of views on the variety of issues is a characteristic of the agile methods [5], [3]. Customer availability and access are also challenging in agile methods [4]. The customer unavailability is mostly because of certain factors like time, cost, and workload of customer representative [24]. This challenge is mostly overcome by having an on-site proxy customer [5] or moving a developer representative to the customer's site [24].

Budget and schedule planning is another challenge. "The cost is usually calculated based on user stories, and new user stories may be included or some may be discarded in forthcoming iterations" [3]. Upfront estimation is not possible when there are volatile requirements and planning [4]. The inappropriate architecture is also a challenge where the architecture finalized by teams in early stage become inappropriate in later stages [4]. Another challenge which can cause massive rework is neglecting non-functional requirements [26]. The agile methods are flexible and allow change, but it can create troubles incorporating this much change and evaluating the consequences of change [6].

The challenges related to RE in agile and some scattered suggestions on how to overcome these challenges can be found in the literature. However, there is a need to develop a cohesive approach to overcome these challenges. Our proposed approach XRE is in line with this need.

XRE is motivated by XP (eXtreme Programming). XP is a light weighted agile methodology that creates working software very fast and with very few defects. XP is based on 12 core programming practices. XP to get maximum benefits emphasizes on using practices at the extreme level. For example, "if testing is good then we will do testing all the time" and pair programming practice is used for code inspection. We identified most valuable RE practices [1] and proposed eXtreme Requirement Engineering (XRE) approach that pushes these practice to extreme.

3   XRE

We, in an earlier work [1], conducted a global survey of RE practitioners as a part of our research on XRE (see figure 1). The aim of the survey was to identify most valuable RE practices as viewed by RE practitioners worldwide. An online questionnaire method was used to identify relative perceived values of RE practices. In total 54 RE experts participated in this survey from twenty-two countries of the world.   These participants included practitioners from different software development organizations that were involved in requirements engineering activities e.g. requirements engineers, requirements analysts, business analysts. We asked practitioners to assess usefulness on a four points scale RE practices. We presented to respondents, 72 RE practices consisting of 66 practice given by Somerville et al. [32] and 6 identified by us from the RE empirical studies reported in [33]. The study aimed at identifying RE practices viewed as most valuable by RE experts.   As a result, we found six RE practices which were perceived as most valuable by majority (>50 percent) of the respondents. Table 1 presents identified practices.

Table 1. High value RE practices

RE Category	Practices
Documentation	Define a standard document structure
Elicitation	Identify and consult system stakeholders
Analysis & Negotiation	Prioritize requirements
Describing Requirements	Use diagrams appropriately
Describing Requirements	Have direct contact with customers to avoid unclear requirements
Management	Uniquely identify each requirement

In this paper, building on our work, we present an RE framework Extreme Requirements Engineering (XRE). Firstly, we show the mapping of high-value practices identified as a result of the global survey to the role of customer representative or product owner (P.O.) in XP/SCRUM. Secondly, report an online discussion with Agile practitioners on the role of P.O. and proxy P.O. and qualitatively analysis of the discussion. Lastly, we propose guidelines that are based on the mapping and findings of the online discussion.

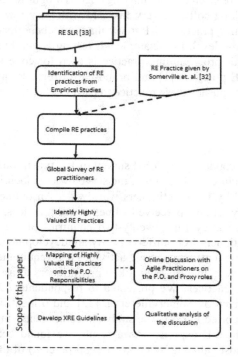

Fig. 1. The XRE research process

4   Mapping of RE Practices on Customer Representative/P.O Responsibilities

Each agile team has an on-site customer representative who is selected by the stake-holders to make decisions on their behalf. The authors [7, 8, 9] of Scrum and XP suggest the customer representative/P.O responsible for conducting most of the RE activities in agile. The customer representative is often called product owner (P.O.) in Scrum. The customer representative or product owner handles defining the product's features, ordering them, reviewing them and accepting or rejecting the results [10].

To acquire a deep knowledge of the standard responsibilities of the P.O. we studied the work by the authors of SCRUM and XP [7, 8, 9]. We developed the list of responsibilities of P.O, and then we mapped the six high value RE practices on these responsibilities. Figure 2 shows the high value RE practices and the matching responsibilities.

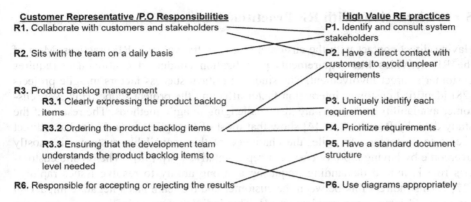

Fig. 2. Mapping of P.O. Responsibilities and High value RE practices

We carried out mapping by asking question "what would it take?" against each of the responsibilities. For example, for R1 we asked, "What would it take to collaborate with customers and stakeholders?" The answer was the identification of customers and stakeholders and working jointly to achieve system/organization goals. From practice perspective, we asked "which responsibility is going to help to carry out the practice?" For example, for P2-Have a direct contact with customers to avoid unclear requirements, we identified two main elements "direct contact" and "avoid unclear requirements". These two elements were then checked against P.O. responsibilities and R1 "Collaborate with ..." and R3.3 "... understands the product backlog items to level needed" were mapped. Mapping of all responsibilities and practices is given in next paragraph.

P1 (identify and consult system stakeholders) is mapped to two responsibilities R1 (collaborate with customers and stakeholders) and R2 (sits with the team on the daily basis), as both of these responsibilities include collaboration with stakeholders (the development team, other customers, users, suppliers, and testers etc.). The practice P2 (have direct contact with customers) is also mapped to two responsibilities these are R1 (collaborate with customers and stakeholders) and R3.3 (Ensuring that the development team understands the product backlog items to the level needed).

Both of these responsibilities R1 and R3.3 are enforcing the communication between the customer and the development team so that the customer (customer representative) can clarify the requirements or backlog items. Hence, the team can accurately implement them.

P3 (uniquely identify each requirement) is mapped to responsibility R3.1 (clearly expressing the product backlog items). Clearly expressing backlog items include giving each backlog item/requirement a unique identifier to avoid ambiguity among different requirements and avoid traceability problems. Practice P4 (prioritize requirements) is mapped to the responsibility R3.2 (ordering the product backlog items). The RE practices P5 (have a standard document structure) is mapped to responsibility R6 (responsible for accepting or rejecting the results). As this is the time when the product owner can accept or reject the results after ensuring that the standards of the organization are properly followed. P6 (use diagram appropriately) R3.3 (Ensuring that the development team understands the product backlog items to the level needed).

5   Discussions with RE Practitioners

Having direct contact with the customer is one of the valuable RE practices. Most of the RE activities like requirements prioritization conducted continuously requires customer's direct involvement. The study on critical success factors in agile projects [28] identified customer interaction factor affecting the project scope. However, customer availability and access are also challenging in agile methods. The results of the study done by Rajesh et. al. [4] show that most of the agile teams do not have direct access to customers. In agile, the challenge of the unavailable customer is mostly overcome by having a proxy customer representative [10], [30]. "The on-site customer's role is indeed demanding, requiring a strong ability to resolve issues rapidly" [27]. This indirect link between the customer and the team has a negative impact on the required intense communication [4]. This indirect communication may cause the risk of decreasing clarity on requirements that in turns may cause more requirements change and need of having detailed specification of requirements [29]. Therefore, it is important to know, what are the issues in using the proxy customer representative instead of real from RE perspective?

To further understand the role of customer representative/P.O in practice we conducted discussions with RE practitioners through LinkedIn groups "Agile" (https://www.linkedin.com/grp/home?gid=81780) and "Lean Agile Software Development Community" (https://www.linkedin.com/grp/home?gid=1024087). These are two most popular groups of agile practitioners. The experienced practitioners involved in agile development participated in the discussions. In "Lean Agile" 12 members participated in our discussion. The roles of these members include Sr. Technical Director, Agile Product Owner & Business analyst, and Scrum coach/master. In "Agile" 11 members participated in the discussion. The roles of these members include Scrum agile expert, Certified Scrum Trainer and Agile Coaches. The focus of discussions was on the role of proxy customer representative in the agile development process in real world setting. We asked, "What is the impact of having a proxy customer representative/PO?" A total of 23 professionals participated in the discussion. All of the participants except one pointed out different issues. To analyze the discussion transcript, we applied the four-stage qualitative data analysis process outlined by [31].

5.1   Stage 1: Collect and Analyze Discussion Data

We carefully inspected each post in the discussion for its research relevance before being including it in the analysis.

5.2   Stage 2: Categorize Data (Coding)

We used coding to identify the concept categories/themes. Coding reduces or groups text into manageable chunks or concept categories. A theme represents the underlying knowledge embedded in the text. For example, we analyzed the statements from discussion and identified that the text (in the transcript of discussion) "How good does your proxy know the business?"· and "Proxy Product Owner should be a subject matter expert" belong to same concept category "Understanding of the business". Table 2 shows the categories and reference number of professionals mentioning it in their responses.

Table 2. Themes identified from discussion with experts

Themes	Expert Ref#	Frequency
Frequent Interaction - with customer and stakeholders	5, 8, 14, 15, 16, 17, 18, 19,20, 21, 22, 23	12
Understanding of the business	2, 3, 5,6,8,9, 12, 13, 17, 21	10
Independent decisions	5,9, 10, 11, 12, 13,14,21	8
Stakeholders Representation	5, 3, 11,17,18, 21	6
Situation dependent	7, 15, 19	3
Business Analyst as proxy works well	1	1
PO should not be too technical	2	1
Amount of time dedicated by proxy	3	1
Losing quality of communication	4	1

We were able to identify four major themes (Table 2 – light pink rows) after the analysis of the discussion transcript. These themes are frequent interaction with customers and stakeholders, understanding of the business, independent decisions, and stakeholders' representation. The participants mostly pointed out the frequent interaction of P.O./C.R. with customers and stakeholders a pre-requisite for success. They were of the opinion that failure to do this will result in lack understanding about business and business value to deliver which was echoed in the responses such as

"How do you maximize the value created by a development team without involving all of the stakeholders?" Certified Scrum Trainer and Coach.

"Alignment needs to be ensured though, as the PO usually is not the (buying) customer. So there is a risk of misalignment which needs to be managed also by getting frequent feedback from real (buying) customers or end users that need the product to provide x value through usable functionality" Enterprise Agile Coach.

The second most referred theme by the participants was understanding the business for which software is being developed. They viewed the lack of understanding created a negative impact on the project.

"The PO can negatively impact the project if he or she does not know enough about the solution/project" Sr. Agile Practitioner.

One of the responsibility of a P.O. prioritization of product backlog items. This activity requires the knowledge of business value as agile development focuses on value driven development and delivery.

The third theme pointed out by the participants was independent decisions. They were of the view that the success of proxy P.O. is dependent on the authority s/he has in making decisions. This was echoed in a comment made by a P.O. of a large company highlighting the link between understanding business and independent decision making.

"What you need in a product owner, is someone who knows about the business enough that they can make independent decisions without having to have meetings with multiple stakeholders every time." Product Owner (at a large company)

The last of the four themes is stakeholder representation. The participants were of the view that proxy P.O. must be a true representative of the customers. One minor theme (in responses of three professionals) is "This depends on the situation". Three of issues were raised by three individuals and thus were not included for themes. These are "Proxy should not be too technical as technical people tend to ignore business", "Proxy must spend most of his/her time full filling this role", and "Having proxy will result in loss of quality communication".

5.3   Stage 3: Apply Thematic Network Analysis

We selected four frequently mentioned concept categories after organizing the data into concept categories.   We applied the thematic network analysis to each concept category to explore the relationship (Figure 3).

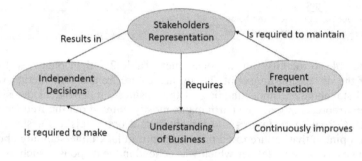

Fig. 3. Thematic Network

5.4   Stage 4: Interpret the Thematic Network Analysis

Finally, we interpreted the thematic network to identify key points. We identified two critical points if an organization decides for a proxy PO/CR.   Firstly PO/CR has to be a true representative that can be visible by the number of independent decisions taken by her/him. Secondly, a true representation of customer requires a good understanding of customer's business and frequent interaction with real customer and stakeholders. These dependencies can be summarized with a help of the comment made by a participant.

"My experience is that as long as the proxy understands the problem and business domain, works with the team and is empowered to make decisions, then the value of having a proxy far outweighs the impact of delays to development when we are waiting for responses from our seldom available PO." Delivery Lead.

6   XRE Guidelines

Agile approaches require an on-site customer representative. However, empirical evidence [4] suggests that most of the teams do not have access to the on-site customer. The challenge of the unavailable customer is mostly overcome by having a proxy customer representative [10], [30]. The agile practitioners, as discussed in the last section, do not view the use of a proxy as a problem as long as proxy representing customer clearly, understand problem and business domain, and can make independent decisions. All of this, as shown in figure 3, requires a frequent interaction of proxy with real customers and stakeholders. This finding from the online discussion is in concurrence with the two highly valued practices (P1 and P2) identified in our earlier research [1] and described in the section 3.

We, on the basis of highly valued RE practice and Agile practitioners' online discussion, propose the XRE framework. The framework consist of six guidelines that we have developed for agile teams. Table 3 shows these guidelines and it also shows the Scrum and XP events where these guidelines should be implemented. In following paragraphs, we provide details of each guideline.

6.1   Guideline 1: (Including Fourth Question in Daily SCRUM)

The daily standup meeting is an event where the activities of development teams are synchronized and a plane for next 24 hour is made. Customer representative/P.O also attends this meeting. In a usual daily standup meeting, the development team members answer three questions:

1. "What did I do yesterday to help the team meet the sprint/iteration goal?"
2. "What will I do today to help the team meet the sprint/iteration goal?"
3. "Do I see any impediment that prevents me or the Development Team from meeting the Sprint/iteration Goal?"

We propose a fourth question for the daily SCRUM or XP meetings that will be answered by the customer representative (real or proxy).

4. "Did you communicate with any of the stakeholders yesterday?"

The main aim of introducing this question is to ensure stakeholders consultation at the extreme level that is on the daily basis. The thematic analysis done in the last section shows that this frequent interaction is required. The customer representative/P.O is responsible for collecting the thoughts and ideas of other stakeholders and sharing them with the development team. The focus of communication can be on specific point like project timeline, cost, requirements, etc. When ensuring the stakeholder's agreement on work done by the development team on the daily basis, we can increase the probability of success of the sprint/iteration.

According to Schwaber et, al. [8] the daily scrum meetings promote a quick decision making and improve knowledge of the development teams. This guideline further increases these two reimbursements given by daily meetings. As the daily standup meetings are short in duration usually 15 minutes so the answer to this question should be very brief.

The customer representative/P.O will wait until the development team is done with answering their questions and do not interfere in their discussion, in the end he will answer his question. This rule is followed because the inventors of SCRUM [11] suggest that no one should ask any question or open any discussion while the development team is reporting their work until the meeting concludes. It is not necessary to meet the stakeholders directly the P.O can use any communication medium. It is not necessary to mention the communication medium while answering the question.

6.2   Guideline 2: (Have a Meeting/Focus Group with Key Stakeholders Before the Sprint Planning)

The customer representative/P.O should conduct a meeting with all interesting stakeholders and discuss the priorities of the product backlog items and understand business value. This should be done through a focus group session as it will help in resolving the conflicts if any. In SCRUM sprint planning meeting or XP iteration kickoff meeting it should be assured that this prioritization meeting was conducted by the customer representative/P.O.

The primary purpose of this meeting is to understand business value and prioritize accurately the backlog items with the involvement of the key stakeholders in the process. With the stakeholders close cooperation we can maintain a clear and prioritized list of demand on the product backlog [10]. The most important responsibility of P.O is to manage needs and expectations of different stakeholders [12]. On-site customer is one of the primary rules and strengths of agile. But some XP and SCRUM project studies [10], [13], [14] reported that they are not able to implement this practice in reality because of problems of customer unavailability and lack of knowledge. In such situations, the practice of on-site customer is implemented by having a knowledgeable engineer or manager play the role of customer who is called a proxy customer. The proxy customer usually does not clearly know the opinion of actual customers and other stakeholders as compared to an actual customer representative. So it is necessary to her/him to discuss different features with them before making any important business decision. We recommend the use of focus group session to conduct this stakeholders' meeting as it is most appropriate and an easy way to facilitate focused communication among groups of stakeholders [15].

Table 3. XRE Practices and SCRUM/XP events

ID	Practices	Scrum /XP Event
G1	Addition of a question in daily standup meeting: Did you communicate with any of the stakeholder yesterday?	Daily Standup Meeting
G2	Have a focus group session with key stakeholders before the sprint planning meeting to discuss business value and prioritization of backlog items	Sprint Planning
G3	Build consensus to ensure that the documents standards/templates are acceptable to all key stakeholders.	Initial/Product Planning Meeting
G4	Check that all documentation is according to agreed standards/templates.	Sprint Review
G5	Check that all the diagrams used are facilitating the communication and understanding among team members.	Sprint Retrospective Meeting
G6	Ensure that each requirement selected from product backlog for a sprint backlog has a unique id.	Sprint Planning

6.3   Guideline 3: (Build Consensus to Ensure that the Documents Standards/Templates are Acceptable to all key Stakeholders)

In agile requirements are documented in the form of user stories. Development teams are often uncertain about what information should be included in the user stories and in what detail [16]. Useful templates are available from many sources that can be used as standards. It is the responsibility of the product owner to develop a consensus over the documentation standard that is accepted by all key stakeholders. "Developing a contract for each user story provides a streamlined mechanism for managers, developers, and testers to agree on details of the story" [17].   These standards can be used later as acceptance criteria for user stories.

6.4   Guideline 4: (Check that all Documentation is According to Agreed Standard)

The agile team must make sure that the document conforms to the agreed standard document structure before accepting the any document including user stories. To check this the team can use any method for example using the standard/template as a checklist. The standard structure for documentation is very important. According to Sommerville et al. [5] by using a standard document structure we can get several benefits such as the use of knowledge from a previous document that results in understanding the relationships among different documents and parts.

6.5   Guideline 5: (Check the Diagrams are used Appropriately)

The agile team should keep checking that the diagrams are used to facilitate the development process and are being used appropriately. In agile software development, different diagrams are used at different levels. For example in Scrum, projects progress is tracked and reported by using the release burn down chart, iteration burn down charts, and a task boards [18]. The UML diagrams can be used to communicate the design within he team. The team members can understand the solution and contribute faster by looking at these diagrams.

In XP before coding short designing sessions are recommended while in FDD (Feature Driven Development) at beginning of each iteration design in executed using UML diagrams [19]. Similarly, the agile DBAs work with application developers and model their needs using UML diagrams e.g. class diagrams [20]. All developers in agile should have basic understanding of industry standard diagrams, especially UML diagrams. It is the responsibility of the agile coach/Scrum mater to keep checking time to time that these diagrams are used to facilitate the development, are in the proper format and are simple enough that everyone can understand.

6.6   Guideline 6: (Check the Every Requirement has a Unique id)

Failure of properly specifying requirements can lead to incorrect, unusable, unverifiable and inconsistent requirements. Traceability problems are one of the major results of not specifying the requirements properly. A clear and unambiguous product back-log contains the requirements with each having a unique identity. "Uniqueness is very important for each requirement to make references to other requirements and build traceability tables" [5]. Also as Cleland-Huang et al. [21] noted that goals of traceability in agile projects are similar to those found in non-Agile projects. When the requirements are entered into the product backlog, these are given a unique identifier. At Sprint/iteration planning meeting when the requirements are selected from the product backlog for the sprint backlog, the P.O should re-check that each requirement has a unique id. Hence, if two requirements are given the same identifier by mistake, it will be removed by the P.O at this stage.

7   Conclusion and Future Work

We, in this paper, presented an agile RE approach, eXtreme Requirements Engineering (XRE) for agile practitioners. This approach is based on RE valuable practices that are taken from our previous study [1] and feedback from the agile practitioners. The XRE consists of guidelines for customer representatives/product owners. By implementing these guidelines, XRE not only ensures the use of valuable RE practices to the extreme but also solves the problem due to a proxy P.O.

Although XRE is based on validated RE practices, it needs to be validated itself. Therefore, our future work consists of the evaluation of our proposed XRE guidelines. The evaluation of XRE will help the practitioners and organizations who want to adopt it and want to know about its significance and importance in real world settings. The evaluation will also indicate the productivity, efficiency, and utility of XRE. We are currently working on conducting focus group study as a first step in the evaluation of XRE.

Acknowledgments. Our thanks to RE professionals for participating in our discussions and sharing their valuable experience with us.

References

1. Ikram, N., Naz, S., Niazi, M.: Valuable Requirements Engineering Practices: An Empirical Investigation. Submitted. In: Requirements Engineering Journal (2015)
2. Zhu, Y.: Requirements Engineering in an Agile Environment: Published master's dissertation Uppsala University, Sweden, Accession No. diva2:233925 (2009)
3. Cao, L., Ramesh, B.: Agile requirements engineering practices: An empirical study: Software. IEEE **25**(1), 60–67 (2008)
4. Ramesh, B., Cao, L., Baskerville, R.: Agile requirements engineering practices and challenges: an empirical study. Information Systems Journal **20**(5), 449–480 (2010)
5. Daneva, M., Van Der Veen, E., Amrit, C., Ghaisas, S., Sikkel, K., Kumar, R., Wieringa, R.: Agile requirements prioritization in large-scale outsourced system projects: An empirical study. In: Journal of systems and software 86(5), 1333–1353 (2013)

6. Inayat, I., Salim, S. S., Marczak, S., Daneva, M., Shamshirband, S.: A systematic literature review on agile requirements engineering practices and challenges. In: Computers in Human Behavior (2014)
7. Pikkarainen, M., Haikara, J., Salo, O., Abrahamsson, P., Still, J.: The impact of agile practices on communication in software development. Empirical Software Engineering **13**(3), 303–337 (2008)
8. Schwaber, K., Sutherland, J.: The definitive guide to Scrum: The rules of the game. www.Scrum.org, https://www.scrum.org/Portals/0/Documents/Scrum%20Guides/Scrum_Guide.pdf
9. Dudziak, T.: eXtreme programming an overview. In: Methoden und Werkzeuge der Softwareproduktion WS (1999)
10. Hoda, R., Noble, J., Marshall, S.: The impact of inadequate customer collaboration on self-organizing Agile teams. Information and Software Technology **53**(5), 521–534 (2011)
11. Sutherland, J., Schwaber, K., Scrum, C.C.O., Sutherl, C.J.: The scrum papers: Nuts, bolts, and origins of an agile process. http://scrumtraininginstitute.com/home/stream-download/scrumpapers. (Last Accessed April 11 2015)
12. Sverrisdottir, H.S., Ingason, H.T., Jonasson, H.I.: The role of the product owner in scrum-comparison between theory and practices. Procedia-Social and Behavioral Sciences **119**, 257–267 (2014)
13. Judy, K. H., Krumins-Beens, I.: Great scrums need great Product owners: Unbounded collaboration and collective Product Ownership. In: Hawaii International Conference on System Sciences, Proceedings of the 41st Annual, pp. 462–462. IEEE (2008)
14. Lohan, G., Lang, M., Conboy, K.: Having a customer focus in agile software development. In: Information Systems Development, pp. 441–453. Springer New York (2011)
15. Stewart, D.W., Shamdasani, P.N.: Focus groups: theory and practice: (Vol. 20). Sage Publications (2014)
16. Rees, M.J.: A feasible user story tool for agile software development? In: Software Engineering Conference, Ninth Asia-Pacific, pp. 22–30. IEEE (2002)
17. Gregorio, D.D.: How the Business Analyst supports and encourages collaboration on agile projects. In: Systems Conference (SysCon), 2012 IEEE International, pp. 1–4. IEEE (2012)
18. Miranda, E., Bourque, P.: Agile monitoring using the line of balance. Journal of Systems and Software **83**(7), 1205–1215 (2010)
19. Alliance, A.: Agile manifesto (2001). http://agilemanifesto.org/. (Last Accessed April 11 2015)
20. Ambler, S.: Agile database techniques: Effective strategies for the agile software developer. John Wiley & Sons Inc, New York, NY (2003)
21. Cleland-Huang, J., Berenbach, B., Clark, S., Settimi, R., Romanova, E.: Best practices for automated traceability. IEEE Computer **40**(6), 27–35 (2007)
22. Wieringa, R.: Software requirements engineering: the need for systems engineering and literacy. Requirements engineering **6**(2), 132–134 (2001)
23. Eisenbarth, M.: Lessons learned from best practice-oriented process improvement in Requirements Engineering–A glance into current industrial RE application. In: Software-technik-Trends 30, no. 1, (2010)
24. Racheva, Z., Daneva, M., Herrmann, A.: A conceptual model of client-driven agile requirements prioritization: Results of a case study. In: ACM-IEEE International Symposium on Empirical Software Engineering and Measurement, p. 39. ACM (2010)
25. Carlson, D., Matuzic, P.: Practical agile requirements engineering. In: proceeding of the 13th Annual Systems Engineering Conference. San Diego, CA (2010)

26. Cardinal, M.: Addressing non-functional requirements with agile practices (2011). http://at2011.agiletour.org/files/AgileTour-Adressing-NFR-with-agile-practices.pdf. (Last Accessed April 11 2015)
27. Koskela, J., Abrahamsson, P.: On-Site customer in an XP project: empirical results from a case study. In: Dingsøyr, T. (ed.) EuroSPI 2004. LNCS, vol. 3281, pp. 1–11. Springer, Heidelberg (2004)
28. Chow, T., Cao, D.-B.: A survey study of critical success factors in agile software projects. Journal of Systems and Software **81**(6), 961–971 (2008)
29. Inayat, I., Salim, S.S., Marczak, S., Daneva, M., Shamshirband, S.: A systematic literature review on agile requirements engineering practices and challenges. In: Computers in human behavior (2014)
30. Bjarnason, E., Wnuk, K., Regnell, B.: A case study on benefits and side-effects of agile practices in large-scale requirements engineering. In: Proceedings of the 1st Workshop on Agile Requirements Engineering (p. 3). ACM (2011)
31. Rene, M., Taylor-Powell, E.: Analyzing qualitative data (2003). http://learningstore.uwex.edu/assets/pdfs/g3658-12.pdf. (Retrieved) (Last Accessed April 11 2015)
32. Sommerville, I., Ransom, J.: An empirical study of industrial requirements engineering process assessment and improvement. ACM Transactions on Software Engineering and Methodology (TOSEM) **14**(1), 85–117 (2005)
33. Ambreen, T.: State of Art in Requirements Engineering: a thesis presented in partial fulfilment of the requirements for the degree of M.S. in Software Engineering at International Islamic University (MS Thesis) (2013)

A Novel Checklist: Comparison of CBR and PBR to Inspect Use Case Specification

Asma Naveed[1,3] and Naveed Ikram[1,2(✉)]

[1] Riphah International University, Islamabad, Pakistan
asmanaveed@fui.edu.pk, naveed.ikram@riphah.edu.pk
[2] Ibn-e-Sina Empirical Software Engineering Lab, Islamabad, Pakistan
[3] Foundation University Islamabad, Rawalpindi, Pakistan

Abstract. High quality and cost effective software development entails early detection of errors from requirement specification artifact/s. For this purpose, various inspection techniques have been presented to identify requirement specification errors. In most reported studies, comparison of two commonly used inspection techniques CBR (Checklist Based Reading) and PBR (Perspective Based Reading) had been conducted to identify defects from the UCS (Use Case Specification); however no comparison was done based on IEEE STD 830-1998 defects' types. Therefore, a novel checklist was developed to identify the IEEE STD 830-1998 specified defects' types namely Ambiguousness, Incorrectness, Inconsistency and Incompleteness from UCS, a major contribution of this research. This developed checklist was later validated to be utilized during this experimental research. In this study, a quasi-experiment was conducted with industrial professionals to compare the effectiveness and efficiency of CBR and PBR using the developed checklist to inspect the UCS that was specified in Use Case 2.0 format. The result of this research showed significant difference between CBR and PBR, i.e. PBR found more defects for all defects' types compared to the CBR technique, but CBR reported less False Positive defects by applying the developed checklist for all defects' types. It was also proved that CBR is more efficient (time based) than PBR for all defects' types. These findings will provide guidelines to industry practitioners for the selection of an inspection technique based on effectiveness, efficiency and false positive ratio for a particular type of defect.

Keywords: Use case specification · Inspection techniques · Defect taxonomies · Checklist based reading · Perspective based reading · Requirement testing

1 Introduction

To remain competitive, the software industry strives to develop high quality products with minimum cost and less development time. However, one major obstacle in attaining such products is late identification of defects. According to Pressman, [1] "…Some maladies, as doctors say, at the beginning are easy to cure but difficult to recognize… but in the course of time… become easy to recognize but difficult to cure". High quality software requires error detection during the requirement phase of software development,

© Springer-Verlag Berlin Heidelberg 2015
L. Liu and M. Aoyama (Eds.): APRES 2015, CCIS 558, pp. 109–125, 2015.
DOI: 10.1007/978-3-662-48634-4_8

because the errors, if not identified in the requirement phase of software development, might propagate to the next phases and result in poor quality and high cost of software product [4].

Various methods for the detection of errors from the requirement specification have been reported in literature e.g. reviews, walkthroughs and inspections [5]. Fagan [6] reported that IBM inspections detected 90% of total defects over the lifecycle of a product, and had a 9% reduction in the average project cost as compared to cases in which walkthroughs were applied. This indicates that quality can be ensured by conducting inspection to identify errors from software requirement specification. The most essential step in inspection is the defects detection phase [7], where the inspectors try to individually identify maximum defects in the document. Therefore, the present research focuses on optimization of this step.

Software requirement are divided into two categories: Functional Requirements FRs and Non-Functional Requirements NFRs [8]. FRs are commonly described as Software Requirement Specification (SRS) and UCS, mostly in the form of UCS [9].The UCS method was initially developed by Jacobson [10]. It is mostly used in the industry to specify requirements due to adoption [9] by the Unified Modeling Language UML [11] which is a de facto standard [12].

Since the errors present in the specification are not known beforehand, it can't be claimed that any inspection technique has detected all the errors in a UCS. However, it can be ensured using the presented checklist that the UCS has been inspected against IEEE specified defects' types.

This significant endowment of this research is a tailored checklist available at: https://www.dropbox.com/s/zoew6as6dbc4ey6/checklist-2.docx?dl=0 {Date visited: August 25, 2015} developed on the basis of defects' types that will assist practitioners in early detection of defects from UCS. It guarantees that inspected UCS has been verified for the presence of quality attributes of Unambiguousness, Correctness, Consistency and Completeness as recommended by IEEE STD 830-1998 [14].

This paper also aims to provide a thorough understanding of the inspection techniques CBR and PBR by comparing them on the basis of defects types from the UCS. Finally, it provides a guideline for the practitioners to choose a better inspection technique in terms of effectiveness and efficiency.

The rest of this paper is divided as follows: Section 2 describes the related work presented earlier, while Sections 3 elucidates experiment design and experiment execution respectively. Section 4 provides a discussion on the results of this research. Sections 5 and 6 present the conclusion and Future Work respectively.

2   Related Work

In previous researches, many experiments reported the comparison between the effectiveness and efficiency of CBR and PBR for the Requirement Specification, and UCS. At the NASA Goddard Space Flight Center, authors empirically proved a remarkable improvement in the effectiveness of the PBR (User, Designer and Tester perspective) in comparison to CBR [15]. A similar result was also reported by a replicated experiment that compared PBR to Ad-hoc and CBR giving similar results [16].

To evaluate the effectiveness and efficiency of PBR (tester perspective) and CBR, a case study was conducted with industrial professionals [17]. The outcome supported PBR (Tester perspective). However, the results of a single case study cannot be generalized. This factor motivated us to conduct an experiment with repeated trials, to compare PBR and CBR in other industrial settings, and prompted us to test the inspection techniques' effectiveness and efficiency for the defects' types in order to improve the overall defects coverage. Another issue in this study was that only the tester perspective was selected while the other perspectives like user and designer also need to be included to achieve higher defect detection rate.

A few experiments, however reported contradictory results. PBR was reported to be less effective as compared to CBR using students as subjects [19][21]. Similarly, in another experiment with students as subjects, PBR (Use Case Analyst, Structured Analyst and Object-Oriented Analyst) again proved to be less effective than CBR and Ad-hoc [20]. The possible reasons for these contradictory results and failure of PBR may be improperly defining the inspectors' role, and providing insufficient instructions to subjects to apply active guidance.

In another experiment, the evaluation of PBR with client and developer perspectives was conducted using UCS as an object of study [22]. The defects were reported for both perspectives of PBR with respect to UCS format such as names of actors and use cases, flow of events, variations and so on. During the inspection, both perspectives used the same checklist. The outcome indicated that inspection effectiveness could be improved by applying different perspectives because different defects were identified by using client and developer perspectives [22].

Previously, a quasi-experiment was performed to examine the impact of active guidance [23]. The checklists and reading scenarios of PBR were designed to be identical, to examine especially the factor of active guidance. It was evident from the outcome that due to the element of active guidance, inspectors with PBR could identify more subtle defects compared to CBR. PBR proved to be more effective but less efficient as compared to CBR because the inspectors had to develop the work product during the inspection process.

Based on the above mentioned literature review, it is established that previous researches did not compare PBR and CBR on the basis of defects' types to ensure maximum coverage of defects detection from the UCS. Therefore, the present research compares CBR and PBR on the basis of defects' types to detect maximum defects from the UCS.

To detect defects during the inspection of requirement artifacts, various defect taxonomies or classifications have been presented in literature. As recommended by the IBM [24], to attain maximum defects detection rate, the inspection technique should be applied by focusing on defect taxonomies which are generic i.e. independent of process or product, so these taxonomies can be applied consistently to any deliverable product like requirement specification, design and code documents. IEEE STD 830-1998 [14] and Neill and Laplante [9] state that the presence of defects i.e. Ambiguousness, Incompleteness, Inconsistency and Incorrectness etc. in requirement specification would have an adverse impact on quality and often require major revisions, or may cause system failure entirely. According to Krogstie [25], syntax and semantics defects can be caused by the absence of quality attributes (Unambiguousness, Correctness, Completeness and Consistency). After analyzing literature for the defect taxonomies regarding SRS, it is concluded that most of the presented defect taxonomies were generic, and applicable to all phases of software development [29]. However, a few reported defects taxonomies were not generic e.g. SRS Structure based defect taxonomy can only be applied to requirement phase, focusing only on Consistency defect type [31]. Consequently, it partially detected syntax and semantics defects.

Similarly, for Use Case Specification, many defect taxonomies for defects detection were reported to be generic [32][33][34]. However in a few studies, the defect taxonomy was not generic (structure specific) e.g. as in [23], so it can only be applied to the requirement phase. Also, most previous reported defect taxonomies considered the syntax and semantic defects [32][34][23] but in some reported studies, the defect taxonomy only focused on syntax defects [33]. After analyzing the aforementioned defect taxonomies it has been concluded that one must consider a generic defect taxonomy that is independent of phase, process and product. Therefore, in this research the defects types of Ambiguity, Incorrectness, Inconsistency and Incompleteness are being used to detect defects from UCS as recommended by IEEE STD 830-1998 [14]. Our defect taxonomy is generic and considers syntax and semantic defects.

Inspection techniques are applied to detect defects in the under inspection artifacts. In the literature, the most referred inspection techniques are CBR [33][34][23], Ad-hoc [29], Scenario Based Reading (SBR) [21], PBR [21][34][22], Usage Based Reading (UBR) [37][38], Technique for Use Case Model Construction and Construction-based Requirements Document Analysis (TUCCA) [40] and Metric Based Reading (MBR) [41]. The comparison of CBR in most reported experiments was carried out with aforementioned inspection techniques. CBR and PBR have been selected for evaluation because during the analysis of previous work, it was concluded that CBR is the most preferred inspection technique in the software industry [39]; and PBR offers multiple roles/perspectives e.g. User, Designer and Tester etc. based inspection [21][34][22][16]. Thus by applying different roles, different defects can be identified from the same document, leading to maximum coverage. SBR does not support role-based inspection whereas UBR [37][38] is similar to the PBR user perspective, except that in UBR the use cases are prioritized by the user and hence only covers one perspective of PBR. Therefore, PBR is preferred over UBR and SBR. The TUCCA [40] technique is not applicable in the present experiment to inspect UCS because it is designed to inspect the requirement document. The MBR [41] has not been applied as it was mentioned by the authors of MBR technique that during their experiment, the inspectors found it difficult to understand, and unlike PBR, the MBR technique would not result in any document/s that can be used during the later software development phases. Therefore, in the present research, the CBR and PBR inspection techniques have been selected for comparison. Both PBR and CBR are Process and Product independent i.e. they can be applied to any artifact and tailored according to any artifact for which inspection has to be carried out [23]. Unlike CBR, which only guides the inspector what to find, PBR guides the inspector how to find it and also provides role-based inspection. Thus it is established that many experiments were conducted to compare the effectiveness and efficiency of CBR and PBR for the defects identification from the UCS but the previous researches did not compare PBR and CBR on the basis of defects' types to ensure maximum coverage of defects detection from the Use Case Specification.

3   Research Methodology

3.1   Experiment Design

A quasi-experiment using Active inspection without meeting [36] was performed to compare the Effectiveness, Efficiency and False Positive Ratio of two inspection techniques CBR and PBR for a particular type of defect in the UCS. The inspection process of experiment is carried out on two different UCS documents, with different subjects/ persons of different organizations / places at different time.

List of Study variables is presented in Table 1.

Table 1. Study Variables

Independent Variables (IV)	Dependent Variable (DV)	Confounding Variables
Inspection techniques (PBR,CBR)	Effectiveness (number of defects)	Object: Controlled UCS Document (complexity approximately same)
	Efficiency (Time needed for the inspection of a UCS in minutes)	Controlled Environment (same working hours, project load, morning duties etc.)
	False Positive Ratio	Controlled Subject (qualification, experience) Uncontrolled (motivation level)

Research Question
Based on the discussion described in section 2, the following question hoists:
    "Is there any difference in CBR and PBR inspection techniques in terms of Effectiveness, Efficiency and False Positive ratio for defects' types in Use Case Specification, or not?"

Hypothesis Development
$H1_0$: There is no significant difference in the ambiguity's number of defects for CBR and PBR.
$H1_1$: There is a significant difference in the ambiguity's number of defects for CBR and PBR.
$H2_0$: There is no significant difference in the ambiguity's efficiency for CBR and PBR.
$H2_1$: There is a significant difference in the ambiguity's efficiency for CBR and PBR.

$H3_0$: There is no significant difference in the ambiguity's false positive ratio for CBR and PBR.

$H3_1$: There is a significant difference in the ambiguity's false positive ratio for CBR and PBR.

Similarly, hypotheses were also developed to evaluate the Effectiveness, Efficiency and False Positive Ratio of other residual defects' types of Incorrectness, Inconsistency and Incompleteness.

Development of Instrument (Checklist)

In CBR, a checklist (list of questions) is provided to inspectors. A survey of the German software industry [36] showed that CBR [6] is the most frequently applied reading technique for requirement inspection. According to a survey of 117 checklists [35], most of them have twenty or more questions, and it suggests that the checklists size should not exceed one page as a longer checklist produces a wrong result due to the cognitive overload of the inspector.

In the past, most of the checklists to inspect SRS were developed by populating the checklist with questions focusing on defects types [17][21]. Similarly, to detect the defects from the UCS, checklists were also used [32][30][33][34][23]. Many of these presented checklists were developed using defect taxonomy based on the Use Case template format but the problem was that some of the detected defects were not clearly related to any particular category of use case format [30]. Due to incorrect allocation of defect into a wrong category, the distribution of defects may be reported wrongly. Consequently, this kind of incorrect reporting creates problems during the correction phase of inspection.

Most of the defects identified by applying the checklist were syntax-related rather than semantic in nature [32]. Whereas, the recommendation [36] for the requirement inspection is that the inspection technique must help to expose more semantic defects in comparison to syntactic defect, because in the later stages of software development the semantic faults are more costly and harder to fix.

A checklist was developed after reviewing theories of text comprehension presented by the discourse analysis community [34] because the use cases are written in the form of structured English. The problem with this approach is that different inspectors may report differently because most of the 7Cs are semantic in nature. Hence, subjectivity is unavoidable. Consequently, the amount of detected defects by applying a checklist would be highly dependent on the ability of the inspectors rather than on the inspection process itself. However, the guidelines recommend that the inspection process should be independent of the inspectors' ability to identify defects. In the past, tailored checklists for three different perspectives i.e. User, Designer and Tester were presented [23]. The questions of the checklists were populated by focusing on some of the subsequent defects types of Incompleteness, Inconsistency, Incorrectness, Hard-to-understand, Over-specified and Unfeasible aspects according to a particular perspective. The drawback of these checklists was that they did not consider questions related to other defects types while developing the checklists for a particular perspective e.g. for User Perspective, they should have also considered the Ambiguity's defect type because one cannot accurately find the defects of Incompleteness and Incorrectness if an ambiguous UCS has to be inspected. Another problem with this checklist is that most of the questions were designed at an abstract level. The whole inspection process is dependent on the checklist/instrument that must be developed with an appropriate level of detail, so that the inspector can understand which defect to find and where to locate it.

Thus, it is concluded that in the past, no checklist was presented on the basis of defects' types to detect defects from the UCS. While using a checklist based on defects' types; it can be assured that inspected UCS has attained maximum coverage of IEEE quality attributes. Therefore in this research, a checklist has been developed on the basis of IEEE defects' types.

To develop this tailored checklist, the earlier presented checklists for the UCSs were reviewed [33][34][22][28], and related questions were sifted out and placed in a new checklist according to a particular defect's type. Additionally, the already-reported effective writing rules of UCS were analyzed by considering the fact that missing an effective writing rule will lead to a particular type of defect in the UCS [26][27][18][13][3]. The questions of checklist are described with details about which defects to find and where to find them in the UCS. Both syntax and semantics related defects can be identified by using developed checklist to detect defects of all types. Also, the length of the checklist for each defect type is kept up to one page to avoid both cognitive overloading and inefficiency of the instrument.

Instrument Validation

Practitioners of the Quality Assurance Departments of software companies B, O and I were requested to review the instrument in order to assess its validity. The inspectors pointed out some redundant questions. The feedback of the pilot study was implemented accordingly i.e. redundant questions were eliminated to improve the efficiency and reliability of the instrument for the defects detection.

Pilot Study

Prior to the execution of the actual experiment, a pilot study was carried out to assess the research design and adequacy of the experimental material. This study was conducted with the help of a Software Development Company P. Three inspectors were nominated by the Quality Assurance Department with similar qualifications and years of professional experience. They were assigned the inspection task randomly. A one hour session was arranged for the inspectors to be acquainted with the inspection process requirements. Three defects of each defect's type were injected to determine the validity and reliability of the instrument/checklist.

The inspectors pointed out some redundant questions in the checklist. They also suggested that some questions must be mentioned with examples for more clarity. Therefore, the presented instrument/checklist was revised based on their feedback.

Experiment Procedure

Sampling

True representative of software industry, software companies E and T have been selected on the basis of high-level maturity of CMMI (Capability Maturity Model Integration) i.e. Company E with CMMI level 3 and Company T at level 5. The experiment was carried out with these 2 companies; 3 participants and 10 UCSs from each company that leads to a total of 60 trails.

Selection of Subjects

The population of the present experimental research is professionals from the software development sector. Three participants with similar qualifications and same years of professional experience were nominated by the Quality Assurance (QA) Department of two Software Companies E and T. These nominated participants also had similar experiences of inspection activities. Afterward, these inspectors were assigned their tasks randomly.

Selection of Object

After the discussion with the software companies, ten use cases were finalized considering the inspectors other project commitments. Companies were requested to provide those documents which have inter-dependent UCSs because they may cause subtle defects e.g. inconsistency defects etc. We put in our best efforts to collect UCSs with almost same complexity level.

Both provided UCSs were written for the database management projects. The validity of document can be judged as the Company E's software is operational in the domain of Hospital Information Management based on this specification. The same criterion is valid for Software Company T's UCSs.

Table 2. Severity Levels of Injected Defects

Level of Severity	Example
Less Severity (Internal to use case)	**Ambiguity:** The Pronoun 'he/she' is used instead of User Subject. **Inconsistency** in sequence number of alternative.
Moderate Severity (Intra use cases)	**Inconsistency:** Use Case 2 and Use Case 3 have the same name i.e. Register a patient.
More Severity (Requirement Specification level)	**Incompleteness:** Missing exceptional flow/ Incomplete precondition. **Incorrectness:** Incorrect post condition

After receiving the document, three defects of each defect type were injected into each Use Case. So overall in both software companies provided UCSs, 120 defects of different severity levels (examples of which are mentioned in Table 2) for each defect type were injected.

3.2   Experiment Execution

The experiment was performed within four consecutive days. The time taken by each inspector differed depending upon the inspection technique and his personal capability. On an average, 2-3 hours per day were spent on the experiment. A total of six participants constituting Group I and II executed the inspection at their own work places for their convenience. Like the pilot study the inspectors from Groups I and II were given separate sessions of approximately one hour about the inspection process. This session also included a question-answer part to resolve the participants' confusions and queries. The inspectors from both companies attended the same session conducted by the same instructor (researcher) to ensure the same understanding level. Additionally, the same material i.e. different procedures to inspect documents with CBR, PBR user and designer perspective, defect injected UCS, different defect logging forms according to defects' types and the same instrument i.e. a checklist, was provided. The participants from both companies were asked to specify the description of the defect for confirmation and evidence. The inspectors were also requested to log/register the total time to identify the defects from a UCS to calculate the efficiency of an inspection technique.

3.3   Validity

Internal Validity
In the present study, best efforts were made in order to control and eliminate threats to internal validity, but threats to internal validity usually cannot be completely avoided. In this experiment, as mentioned earlier, the selection of the software companies was not random but based on the maturity level and the availability of requirements documentation in the UCS format. Similarly, the selection of inspectors was not random but was done by the software company itself considering the availability and relevance with the area of requirement testing. To avoid threats to internal validity, the allocation of inspection techniques was done randomly within the delegated teams from both companies.

The inspectors were given ten different UCSs with different injected defects, to eliminate the threats of the learning effect that can influence the internal validity of the experiment. Another threat to the internal validity was the previous knowledge of the participants, which was controlled in the present experiment by the selection of inspectors/participants having almost the same qualifications and years of experience in the field of requirement testing. For the participants' convenience, the study was conducted in their companies and they were requested to avoid sharing the experiment's outcome with each other. Also, as mentioned earlier, the UCS document with the same seeded defects, same instrument i.e. checklist and same training by the same resource person was provided to avoid threats to internal validity.

External Validity
In this research, the external validity is achieved by using an independent design of the experiment. During the experiment, the sample of objects and subjects was taken from two different software companies. So the inspection process of experiment was

carried out on two different UCSs documents with almost the same complexity, but with different subjects/persons of different organizations/places at different times. The reporting of the defects was done by using the self-inventory method of test administration. The participants were asked to specify the description of the defect for confirmation and evidence. By analyzing the description of defects, it can be identified that the detected defects are actual defects and that the inspectors had followed the inspection process as required. The confirmation that the inspection process was rightly applied can also be done by participants/inspectors of experiments who were assigned inspection technique PBR (user and designer perspective), since they write the UCSs and draw the state diagrams respectively, as a procedural step.

Conclusion Validity

To avoid threats to the conclusion validity, the received results, after the execution of the experiment were compiled and carefully checked. The reliability and understandability of the instrument is essential for the conclusion validity, therefore after the execution of the pilot study, redundant questions were eliminated. To remove the threats of "poor reliability of treatment implementation" [48], a self-inventory of defect logging with question numbers along with the description of detected defects was carried out. In addition to this, reliability of treatment implementation was confirmed by artifacts of UCSs and state diagrams developed as a part of PBR inspection technique. The participants were also asked to conduct the experiment in their own work environment for their convenience. In this way, external disturbances were avoided to eliminate the threats of "random irrelevancies in the settings" [2]. Also, by considering 0.05 as a value of significance, credibility of the results was assured.

Construct Validity

The practitioners of QA Department of Software Companies (B, O, P and I) carried out the validation of construct/instrument/checklist. The instrument was later revised by implementing their recommendation.

4   Results and Analysis

4.1   Data Preparation

The Inspectors-submitted results of each inspection technique were matched with the seeded defects. The inspectors also detected unseeded defects. These unseeded defects were separated and analyzed with the help of a domain expert to determine if they were true or false positive defects. The final defects for the PBR were obtained by taking the union of detected defects for the user and designer perspectives of PBR in the case of Seeded, Unseeded and False Positive defects as a procedural requirement of inspection technique. While calculating the efficiency of the PBR inspection technique, time taken by both perspectives was compared and the PBR perspective that consumes more time for defect detection was considered.

4.2   Data Analysis

The objective of this study is to compare the Effectiveness, Efficiency and False Positive Ratio between CBR and PBR on the basis of defects' types. To compare the defect detection rate (effectiveness) of CBR and PBR for the defects' types, 30 defects of each defect's type were seeded in each UCS.

Table 3. Comparison of CBR and PBR for Seeded, Unseeded True and False Positive Defects for companies E and T

			Ambiguity	Incorrectness	Inconsistency	Incom-pleteness
Seeded Defects	Total Seeded Defects for both companies E and T		30	30	30	30
	Company E	CBR	19	18	14	13
		PBR	23	28	26	13
	Company T	PBR	21	25	27	19
		CBR	20	24	24	13
Unseeded Defects	Total True Un-seeded Defects for Company E		61	52	32	100
	Company E	CBR	47	28	16	39
		PBR	54	43	30	85
	Total True Un-seeded Defects for Company T		26	14	14	35
	Company T	PBR	26	14	14	35
		CBR	12	7	5	22
False Positive Defects	Total False Positive Defects for Company E		58	83	45	57
	Company E	CBR	29	18	15	19
		PBR	42	72	41	50
	Total False Positive Defects for Company T		18	28	20	30
	Company T	PBR	18	28	20	30
		CBR	7	16	6	13

The results of Table 3 show that PBR found more Seeded defects for both companies E and T.

The Unseeded True defects identified by experts for all defects' types were checked against inspectors' reported CBR and PBR results as shown in above Table 3. A noteworthy verdict is that PBR reported more defects than CBR for all defect types and the PBR detected defects included all the CBR detected defects besides other detected defects. Interestingly, the number of Incompleteness type of defects is very high. A possible reason for this increase in Incompleteness defect' type may be Incorrectness and Inconsistency types of defects which were also interpreted by inspectors as Incompleteness' type of defects.

The false positive defects identified by expert for all defects' types were checked against inspectors' reported CBR and PBR results. Compiled results shown in Table 3 also indicate that PBR identified more false positive defects as compared to CBR which is not a supportive argument for an inspection technique. The possible reason for PBR detecting more false positive defects is that as a part of required procedure of PBR, two inspectors with different perspectives were inspecting the UCSs.

Table 4. Effectiveness of Inspection techniques for Companies E and T

		Ambiguity	Incorrectness	Inconsistency	Incompleteness
Company E	CBR	73%	56%	48%	40%
	PBR	85%	87%	90%	75%
Company T	CBR	57%	70%	66%	54%
	PBR	84%	89%	93%	83%

During the analysis of results, effectiveness was calculated for CBR and PBR separately. The results of Table 4 show that PBR is more effective than CBR to identify all types of defects.

Table 5. Efficiency of Inspection Techniques for Companies E and T

		Ambiguity	Incorrectness	Inconsistency	Incompleteness
Company E	CBR	55	51	25	43
	PBR	22	20	12	18
Company T	CBR	38	34	33	43
	PBR	21	17	18	21

Efficiency was calculated as number of true defects identified per hour, listed in Table 5. It can be concluded from results that PBR is less efficient. The reason may be that the inspectors applying the PBR need more time because they have to develop artifacts and model too.

Table 6. False positive defect ratio for Companies E and T

		Ambiguity	Incorrectness	Inconsistency	Incompleteness
Company E	CBR	50%	22%	33%	33%
	PBR	72%	87%	91%	88%
Company T	CBR	39%	57%	30%	43%
	PBR	100%	100%	100%	100%

False positive ratio was calculated as: (false positive defects identified by inspection technique) ÷ (expert's reported false positive defects) presented in Table 6. Both companies' presented results indicated more false positive defects ratio for PBR. Moreover, PBR reported 100% false positive defects for all defects' types in company T because PBR found same defects which were also found by CBR, in addition to other defects.

Analysis on the Basis of Statistical Results

For the generalization of results, statistical analysis was performed. The difference between effectiveness of CBR and PBR is found by applying non-parametric chi-square test at 1 degree of freedom as our experiment design has one factor (inspection technique) with two treatments (CBR and PBR) and the data of defects' types is nominal. The chi-square results for the Seeded defects, Unseeded defects, Overall true defects and False Positive defects are presented in below Table 7. Whereas Parametric t-test is applied to find the difference between efficiency of CBR and PBR on the basis of time spent in detecting defects presented in Table 8, as experiment design has one factor (inspection technique) with two treatments (CBR and PBR), the data of time spent in detecting defects is ratio and also distribution of data is normal.

Table 7. Overall Comparison of CBR and PBR effectiveness for Companies E and T for developed Checklist

Defects' types	Chi-Square	Seeded Defects			Unseeded True defects			Overall True Defects			False positive Defects		
		Company E	Company T	Better inspection Technique	Company E	Company T	Better inspection Technique	Company E	Company T	Better inspection Technique	Company E	Company T	Better Inspection Technique
Ambiguity	χ^2	1.27	.077	PBR	2.82	19.2	PBR	3.95	20.35	PBR	6.14	6.76	CBR
	df	1	1		1	1		1	1		1	1	
	p	.260	.781		.093	.000		.047	.002		.013	.009	
Incorrectness	χ^2	9.32	.111	PBR	9.99	9.33	PBR	18.7	4.47	PBR	70.77	6.87	CBR
	df	1	1		1	1		1	1		1	1	
	p	.002	.739		.002	.002		.000	.034		.000	.009	
Inconsistency	χ^2	10.8	1.176	PBR	15.15	13.3	PBR	25.7	10.06	PBR	31.95	12.2	CBR
	df	1	1		1	1		1	1		1	1	
	p	.001	.278		.000	.000		.000	.002		.000	.000	
Incompleteness	χ^2	.000	2.41	PBR	44.91	15.9	PBR	33.3	12.86	PBR	35.23	10.0	CBR
	df	1	1		1	1		1	1		1	1	
	p	1	.121		.000	.000		.000	.000		.000	.002	

Chi-square test values for the seeded defects of both companies E and T to compare the effectiveness of CBR and PBR are presented in Table 7. The Null hypothesis is rejected at (0.05) level of significance for two defects' types of Incorrectness and Inconsistency (p=.002 and p=.001) of company E because the calculated values of chi-square are less than 0.05. Therefore, it is concluded that there is significant difference in the effectiveness of CBR and PBR for the said defects' types. However, for Ambiguity and Incompleteness defects' types of company E and for all defects' types of company T, we accept Null hypothesis at (0.05) level of significance as the computed value of p are greater than 0.05 (p>0.05). Therefore, it is concluded that there is no difference in PBR and CBR. The possible reason may be that Company T is at CMMI level 5, so the inspectors from these companies are more experienced than those of Company E. The inspector has also detected Unseeded True Defects other than seeded defects for all defects types, we reject the Null Hypothesis at 0.05 level of significance as the (p<0.05) except for ambiguity defect type.

Thus for Unseeded True defects, there is difference in effectiveness of CBR and PBR; PBR proved to be more effective for these defects. To compare effectiveness of CBR and PBR for the Overall true defects for both companies E and T, chi-square results were computed that rejected the null hypothesis at (0.05) level of significance for all defects' types because the calculated values of p are less than 0.05. Therefore, it is concluded that there is a significant difference in the effectiveness (true defects) of inspection techniques for all defects' types. PBR detected more defects of each particular defect's type in comparison to CBR. The reason for this difference may be that during execution of active inspection technique PBR, the inspector gets better understanding of defects by developing artifacts that is a part of its inspection procedure and also by inspecting it with different perspective resulting in more defects.

The chi-square values of false positive defects show significant difference in PBR and CBR for every defects' types at (0.05) level of significance as (p<0.05). It can been seen that PBR identified more false positive defects as compared to CBR, which is not a supportive argument for an inspection technique. The possible reason is that as a part of procedure, two inspectors with different perspectives were inspecting the UCSs. Thus, the inspection by two inspectors of PBR resulted in reporting of more false positive defects as compared to CBR with one inspector.

Table 8. Comparison of CBR and PBR Efficiency for Companies E and   T

Defects' Types	Inspection Techniques		T-test(p-values)	
	PBR	CBR	Company E	Company T
Ambiguity	👎	👍	8.06E-05	6.12E-14
Incorrectness	👎	👍	1.24E-05	1.02E-14
Inconsistency	👎	👍	1.13E-06	2.78E-12
Incompleteness	👎	👍	0.0010	1.59E-16

The time difference between CBR and PBR to detect a particular type of defect is described in above mentioned TABLE 8. On the basis of t-test results, the null hypothesis is rejected at (0.05) level of significance for all defects' types because the computed values of t-test are less than 0.05. Therefore, it is concluded that CBR is more efficient than PBR for detection of all defects' types. The reason for this is that each inspector of PBR (user, designer) spent more time in comparison of CBR inspector to detect defects because they have to develop model as a part of inspection procedure. The result of this study confirms the past reported studies' results [16][23] where the CBR proved to be more efficient than PBR.

5   Conclusion and Contribution

Based on the Results and Analysis presented in Section 4, we conclude that PBR detected more true defects against seeded as well as unseeded defects. However, for applying PBR, the effort (person hours) as a procedural demand, is more than CBR. CBR is better than PBR for the aspects of efficiency and false positive ratio. We also calculated the ratio of true and false defects to find comparison of actual effectiveness for both inspection techniques. It proved that CBR found more true defects against false positive defects. Therefore CBR is recommended to be used in industry because it is more efficient, reports less false positive defects, has a higher true against false ratio and also needs fewer resources (person hours) in comparison to PBR. The outcome provides the guidelines below to practitioners. Practitioners can now apply our novel developed checklist to detect the defects during early phase of software development.

- For medium to small companies, where resources and affordability are limited, CBR is recommended for inspection. However, it is suggested that in companies where the resources are sufficient, PBR is a better choice for inspection as it finds more true defects (seeded and unseeded) but it consumes more resources in terms of effort (person hours)
- Another significant finding for the industry is that the choice of both inspection techniques is not affected by defects' types since they are not sensitive towards defects' type.
- Results also guides that the practitioners must observe care while authoring and inspecting the UCSs for the Incompleteness and Ambiguity defects' types because their proportion is very high as compared to other defects' types.
- In addition, the result of both companies E and T indicates that the PBR inspectors reported less overlapping as compared to the CBR inspectors.

6   Future Work

Due to limitation of resources like practitioners' time, the scope of this research was only restricted to the comparison of two inspection techniques CBR and PBR. In the future, the other inspection techniques can be compared for guiding the software industry in finding the most effective and efficient inspection technique based on defects type in the UCS. As it was found that both PBR and CBR are not sensitive to defects types, it is to be checked in future if other inspection techniques are sensitive to defects' types or not. PBR can be analyzed by increasing the number of perspectives to find the increase in number of defects. This will guide the industry about the optimal level of perspective for optimal coverage of defects.

Due to prevalence of Agile software development ASD, the new tailored checklist on the basis of defects' types can be developed for user stories and the same experiment to compare the effectiveness and efficiency of PBR and CBR may be conducted to assist the industry to provide maximum defects detection at early phase of ASD in order to ensure high quality of software products in a cost effective way.

Acknowledgment. Thanks to Barbara Peach, for her guidance on the checklist. We also thank all the industry professionals who gave their precious time to complete this experiment.

References

1. Pressman, R.S.: Software Engineering. system **5**, 47B (1999)
2. Trochim, W.M., Donnelly, J.P.: Research Methods Knowledge Base. Atomic Dog /Cengage Learning, Mason, OH (2008)
3. Rolland, C., Achour, C.B.: Guiding the construction of textual use case specifications. Data Knowl. Eng. **25**(1), 125–160 (1998)
4. Boehm, B., Basili, V.R.: Software Defect Reduction Top 10 List, Softw. Eng. Barry W Boehms Lifetime Contrib. Softw. Dev. Manag. Res., 69(1), 75 (2007)
5. Aurum, A., Petersson, H., Wohlin, C.: State-of-the-art: software inspections after 25 years. Softw. Test. Verification Reliab. **12**(3), 133–154 (2002)
6. Fagan, M.E.: Advances in software inspections. In: Pioneers and Their Contributions to Software Engineering, Springer, pp. 335–360 (2001)
7. Gilb, T., Graham, D., Finzi, S.: Software inspection. Addison-Wesley Longman Publishing Co., Inc. (1993)
8. Nuseibeh, B., Easterbrook, S.: Requirements engineering: a roadmap. In: Proceedings of the Conference on the Future of Software Engineering, pp. 35–46 (2000)
9. Neill, C.J., Laplante, P.A.: Requirements engineering: the state of the practice. Softw. IEEE **20**(6), 40–45 (2003)
10. Jacobson, I.: Object-oriented software engineering: a use case driven approach. Pearson Education India (1992)
11. Rumbaugh, J., Jacobson, I., Booch, G.: The Unified Modeling Language Reference Manual 1998, Harlow Addison-Wesley (1990)
12. Lange, C.F., Chaudron, M.R.: Effects of defects in UML models: an experimental investigation. In: Proceedings of the 28th international conference on Software engineering, pp. 401–411 (2006)
13. Somé, S.S.: Supporting use case based requirements engineering. Inf. Softw. Technol. **48**(1), 43–58 (2006)
14. I.C.S.S.E.S. Committee, I. Electronics Engineers, and I.-S. S. Board: IEEE recommended practice for software requirements specifications: approved June 25 1998, vol. 830. IEEE (1998)
15. Basili, V.R., Green, S., Laitenberger, O., Lanubile, F., Shull, F., Sørumgård, S., Zelkowitz, M.V.: The empirical investigation of perspective-based reading. Empir. Softw. Eng. **1**(2), 133–164 (1996)
16. Ciolkowski, M.: S.E.G.S. mit Generischen, Empirical investigation of perspective-based reading: A replicated experiment. Fraunhofer-IESE (1997)
17. Berling, T., Runeson, P.: Evaluation of a perspective based review method applied in an industrial setting, IEE Proc.-Softw., 150(3), 177–184 (2003)
18. Cockburn, A.: Writing effective use cases, vol. 1. Addison-Wesley Boston (2001)
19. Lanubile, F., Mallardo, T., Calefato, F., Denger, C., Ciolkowski, M.: Assessing the impact of active guidance for defect detection: a replicated experiment, In: Proceedings of the 10th International Symposium on Software Metrics 2004, pp. 269–278 (2004)
20. Lanubile, F., Visaggio, G.: Evaluating defect detection techniques for software requirements inspections, ISERN Rep. No 00-08 (2000)
21. Halling, M., Biffl, S., Grechenig, T., Kohle, M.: Using reading techniques to focus inspection performance, In: Proceedings of the 27th Euromicro Conference 2001, pp. 248–257 (2001)

22. Anda, B., Hansen, K., Sand, G.: An investigation of use case quality in a large safety-critical software development project. Inf. Softw. Technol. **51**(12), 1699–1711 (2009)
23. Denger, C., Ciolkowski, M., Lanubile, F.: Does active guidance improve software inspections? A preliminary empirical study. In: Proceedings of the IASTED International Conference on Software Engineering (SE) (2004)
24. Chillarege, R., Bhandari, I.S., Chaar, J.K., Halliday, M.J., Moebus, D.S., Ray, B.K., Wong, M.-Y.: Orthogonal defect classification-a concept for in-process measurements. IEEE Trans. Softw. Eng. **18**(11), 943–956 (1992)
25. Krogstie, J.: A semiotic approach to quality in requirements specifications. In: Organizational Semiotics, Springer, pp. 231–249 (2002)
26. Adolph, U.S., Bramble, P., Cockburn, A., Pols, A.: Patterns for effective use cases. Addison-Wesley Professional (2002)
27. Achour, C.B., Rolland, C., Maiden, N.A.M., Souveyet, C.: Guiding use case authoring: Results of an empirical study. In: Proceedings IEEE International Symposium on Requirements Engineering 1999, pp. 36–43 (1999)
28. Denger, C., Paech, B.: An Integrated Quality Assurance Approach for Use Case Based Requirements. In: Modellierung, pp. 59–74 (2004)
29. Sandahl, K., Blomkvist, O., Karlsson, J., Krysander, C., Lindvall, M., Ohlsson, N.: An extended replication of an experiment for assessing methods for software requirements inspections. Empir. Softw. Eng. **3**(4), 327–354 (1998)
30. Shull, F., Rus, I., Basili, V.: How perspective-based reading can improve requirements inspections. Computer **33**(7), 73–79 (2000)
31. Halling, M., Biffl, S., Grechenig, T., Kohle, M.: Using reading techniques to focus inspection performance. In: Proceedings of the 27th Euromicro Conference 2001, pp. 248–257 (2001)
32. Anda, B., Sjøberg, D.I.: Towards an inspection technique for use case models. In: Proceedings of the 14th international conference on Software engineering and knowledge engineering, pp. 127–134 (2002)
33. Cox, K., Aurum, A., Jeffery, R.: An experiment in inspecting the quality of use case descriptions. J. Res. Pract. Inf. Technol. **36**(4), 211–229 (2004)
34. Phalp, K.T., Vincent, J., Cox, K.: Assessing the quality of use case descriptions. Softw. Qual. J. **15**(1), 69–97 (2007)
35. Brykczynski, B.: A survey of software inspection checklists. ACM SIGSOFT Softw. Eng. Notes **24**(1), 82 (1999)
36. Laitenberger, O., Vegas, S., Ciolkowoski, M.: The state of the practice of review and inspection technologies in germany, Tech Report Number: ViSEK/011/E (2002)
37. Thelin, T., Runeson, P., Wohlin, C.: An experimental comparison of usage-based and checklist-based reading. IEEE Trans. Softw. Eng. **29**(8), 687–704 (2003a)
38. Thelin, T., Runeson, P., Wohlin, C.: Prioritized use cases as a vehicle for software inspections. Softw. IEEE **20**(4), 30–33 (2003b)
39. Fogelström, N.D., Gorschek, T.: Test-case driven versus checklist-based inspections of software requirements–an experimental evaluation. In: WER07-Workshop em Engenharia de Requisitos, pp. 116–126. Toronto, Canada (2007)
40. Belgamo, A., Fabbri, S., Maldonado, J.C.: TUCCA: improving the effectiveness of use case construction and requirement analysis. In: International Symposium on Empirical Software Engineering 2005, p. 10 (2005)
41. Bernardez, B., Genero, M., Duran, A., Toro, M.: A controlled experiment for evaluating a metric-based reading technique for requirements inspection. In: Proceedings of the 10th International Symposium on Software Metrics 2004, pp. 257–268 (2004)

A Kind of Safety Requirements Description Method of the Embedded Software Based on Ontology

Fengjie Zhan, Xiaoyu Wang, Huaxiao Liu(✉), and Lei Liu

College of Computer Science and Technology, Jilin University, Changchun 130012, China
liuhuaxiao@jlu.edu.cn

Abstract. With the popularization of the application of the embedded system, the embedded software has been developed rapidly and then becomes a new growing power. So the research on the safety requirements of the embedded software is becoming more and more important. However, for the analysis of safety requirements, a very important step is the description of safety requirements. This paper proposes a description method of the embedded software based on ontology. And ontology is good for sharing and reuse of knowledge, as well as automatic reasoning. For the embedded software, we use ontology to define the concepts and their relations, and then we construct a model to describe the common concepts. Finally, we give an example to prove our model useful.

Keywords: Ontology · The embedded software · Safety requirements

1   Introduction

Software safety requirements [1-4] are non-functional requirements, emphasizing what the system should do to eliminate software safety flaws [5]. These requirements aim at preventing the assets from accident harm, detecting and responding to the safety events [7-8]. So they should be described as fully as possible and must be clear and accurate 6. At present, most enterprises describe the safety requirements by using natural language, which means that it will cause ambiguity. In order to tackle this problem, many researchers are committed to use different description methods to model safety requirements:

Li Bo [9] proposed a modeling method of software safety requirements based on UMLsec, which can increase the safety related information for model elements by stereotype and describe basic safety requirements(such as confidentiality, integrity, etc.) accordingly.Zhe Chen and Giles Motet [10] used controlling automata to describe safety requirements. They considered safety requirements as control structures that restrict system behaviors at meta-model level. Then they built two models,called system behavior model and system safety constraints model. With these two models combining together, a safety system was established.

Kalloniatis [11] put forward a method based on scenarios to analyze the cloud deployment safety model and set up a requirement model about cloud safety. Then this method helped Greek National Gazette finish the analysis on the safety of the cloud deployment model.

© Springer-Verlag Berlin Heidelberg 2015
L. Liu and M. Aoyama (Eds.): APRES 2015, CCIS 558, pp. 126–134, 2015.
DOI: 10.1007/978-3-662-48634-4_9

In this paper, we use ontology to describe software safety requirements. This method is based on the idea of active defense. Ontology is the philosophical study of the nature of being, becoming, existence, or reality, as well as the basic categories of being and their relations. So it can solve the problem of knowledge sharing and reuse [12] effectively. Ontology is mainly used in the model phase and validation phase of software safety requirements. Because the information in the ontology can answer the relationships among objects within a certain domain, express the domain knowledge and the domain axiom and form validation rules of software safety requirements, ontology can support the validation of software safety requirements during software engineering. Besides, because ontology can be reused, it avoids repeatability of domain knowledge analysis [13-16].

This paper is organized as follows: Section 2 defines software safety requirements description model based on ontology. To analysis and verify the viability of our method, we give an example in Section 3. Finally conclusions are presented in Section 4.

2 Safety Requirements of the Embedded Software Based on Ontology

There are many features in the embedded system, such as small volume, strong performance, low power, high reliability, making the embedded system play a vital role in the fields of the control of process industry, national defense and military, electronic equipment, modern weapons, network communication equipment and consumer digital products. This illustrates that embedded system has a broad market demand and prospects for development. At the same time, the connectivity of the network has greatly changed the traditional definition of embedded system. From the handheld device to the medical monitor, industrial controller and the large telephone system, it is difficult to find an embedded system that is not connected to the network. Moreover, the automotive safety standard ISO 26262, which represents an application-specific refinement of the well-established standard IEC 61580, also presents the requirements for the development of embedded software. So that safety requirements is very important for embedded system, which also is one of the main concerns for the designers and developers.

Embedded software are constantly developing, its safety requirements also followed constantly changing. So, in the long term, using ontology to describe the safety requirements of embedded software is much better than any other description methods. Because ontology can capture knowledge in related fields, provide a common understanding of knowledge, determine the common recognition of vocabulary and define these terms and their relations from different levels of formal model. It means that to construct ontology can achieve knowledge sharing and reuse, and improve the ability of communication, interoperability, reliability. Thus, the main work of this paper is to describe the embedded system safety requirements using the method based on ontology.

Requirements description method based on ontology focus on the concepts and the relations between these concepts. These concepts rely on the real knowledge system that the definer wants to define. According to the standard ISO 26262 and the characteristic of embedded software, firstly, we define concepts of safety requirements of embedded software, and confirm the relations among these concepts in order to get ontology about safety requirements of embedded software. Then we determined whether our ontology is too simple or too complex by manually watching. If it is too complex, we will cut down some redundant or unnecessary description; if it is too simple, we will concrete some concepts further. By repeating above operation, we finally have the following definitions of software safety requirements description model basing on ontology in this paper:

Definition 2.1 Safety Requirements Description Model Of the Embedded Software Based On Ontology (SSM): SSM is defined as a three tuple that is *SSM = (Con, R, Res)*, where:

- *Con*: the safety requirements concepts in the SSM, $Con=\{rC, cC\}$ and $rC \supset cC$. rC is the collection   of root safety requirements concepts, $rC=$ SP∪HP∪ENV, and root safety requirements concepts contains the software part, hardware part and system environment part; cC is the collection of child safety requirements concepts, including a series of specific safety   requirements in the root requirements concepts. As Table 1 show.

- *R:* the relations collection in SSM, $R=rR\cup cR\cup bR$. rR is the relations between the rCs; cR is the relations between the cCs; bR *is* the relations between the rCs and cCs.

- *Res*: set of constraints in SSM.

According to the definitions of SSM, we construct the relations of the concepts of embedded software in Fig. 1

Table 1. The concepts in SSM

concept	meaning
user	The one who use the system service
authentication	Identify one's authority
network_access	Access to the network
communication	Communicate with others
storage_sys	Store user's information and log file
operation	It is the collection of subcommand that supplied by the software or hardware system and it can be carried on when the system is running normally or abnormally
condition	The environment in which the system is running
sys_behavior	System's behavior
hazard	The conditions that result in software failure, e.g. insulin overdose(the case in section 3)
leak	Caused by software risks and can be used for attacking the software safety bugs, e.g.buffer overflow
success	Fault tolerant success and the hazard can be ignored
failure	Fault tolerant failure and the hazard may result in safety failure.
fault tolerance mechanism	The methods on which systems adjust themselves to avoid the occurring of software safety accident while safety risk happens. e.g. file allocation table
safety failure	The safety accidents that already happened or the safety accidents that must happen because of no measures, e.g. firewall is destroyed

Table 2. The relations in SSM

Relation	Format	meaning	example
contain	rC->cC OR rR->rR	Represents the inclusion relation between cC and rC or cC and cC	operation->network_access operation->communication
stored_in	cC->cC	Represents the left cC can store in the right cC	user->storage_sys
control	cC->cC	Represents the left cC can control the right cC	fault tolerance mechanism->hazard
result_in	cC->cC	Represents the left cC can lead to the right cC	sys_behavior->hazard
use	cC->cC	Represents the left cC use the right cC	user->authentication
author - ized_by	cC->cC	Represents the left cC grant authorization to the right cC	operation->user

Modeling object of this article is cC, which composes rC. Because accurate description of cC can determine the relations among rC and also guide engineers to model well in the other phase of software engineering. And Table 1 and Table 2 have given the concepts and relations of the embedded system that we need to define according to SSM.

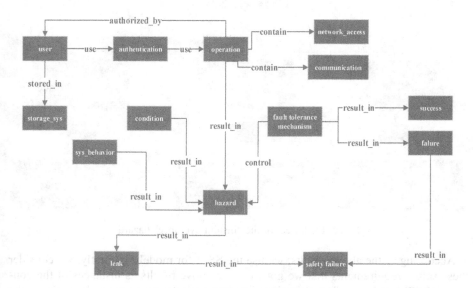

Fig. 1. The relations and concepts in SSM

We define and implement the constraint set Res in SSM, which is combining with the characteristics of the safety requirements, and obtained by summarizing the existing requirements model integrity and consistency. In this model, the software safety requirements constraints are bound to have:

Res1: Belonging Constrains: every cC must belong to an rC, but may not belong to only one rC.

Res2: Containing Constrains: Every rC contains one cC at least.

Res3: Uniqueness Constrains: The name of all concepts in Con must be unique and unrepeatable.

Res4: Relation Constrains: All R must be a certain relation between Con and Con.

Res5: Existence constraints: All the concepts in Con must be able to obtain and define from the actual needs of the safety needs of the software.

3   Case Analysis

In this section, we will use our modeling method to model and analyze a case to show the applicability of this method. Taking the insulin pump as an example, Luiz Eduardo et al analyze the excess insulin injection of insulin pump failure [18-19] with the fault tree analysis methods, and get the safety requirements, such as shown in Fig. 2 [17].

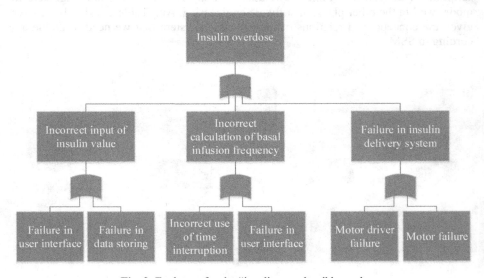

Fig. 2. Fault tree for the "insulin overdose" hazard

According to the above results, we use the SSM for modeling. Firstly, we consider these safety requirements that we got from the above results as instances of the concepts of SSM and classify the requirements into the relevant concepts by defining the concepts in software safety requirements, the results are shown in Table 3.

Table 3. Concepts Table

Concept	Entity
operation	input of insulin value
	calculation of basal infusion frequency
	use of time interruption
user	user interface
storing_sys	data storing
sys_behavior	time interruption
hazard	insulin overdose

Next, we put the software safety-critical requirements concepts entity into SSM, and determine the relation in part of software safety requirements .We need to determine the relation between two entities and take the relations between user and operation as example to describe. The result is shown in Fig.3.

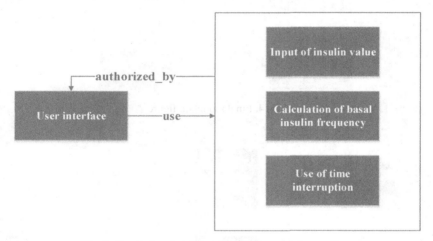

Fig. 3. Concept and relation figure of user and operation

Then, we need to combine every two software safety requirements' figure and get the final software safety requirements. Using the software safety requirements description method of this paper, the description of the requirements is finally obtained as shown in Fig.3.

Comparing Fig.3 with Fig.4, we find that the extraction of the insulin infusion pump's software safety requirements is insufficient. Luiz Eduardo's extraction of safety requirements without considering the environmental factors, fault tolerance mechanism and fault tolerance mechanism failure which may lead to safety failure problems and risks that may lead to vulnerabilities. Add these issues in the Fig.3, we get Fig.4. The oval parts in the Fig.4 are the concepts, which need safety engineers to find out relevant entities about "insulin overdose" to fill in. As is shown in the Fig.4, we can see clearly the relationship among these requirements. This model can help software engineers find out that what kind of operations or system behaviors can result in "insulin overdose" and what kind of fault tolerance can control this issue immediately. It can also tell engineers what kind of issues they should consider to prevent insulin overdose during the software design phase.

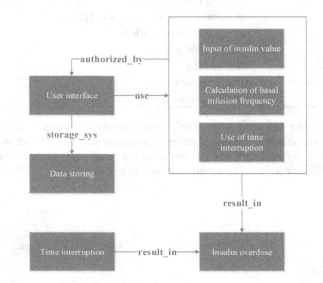

Fig. 4. Final modeling figure

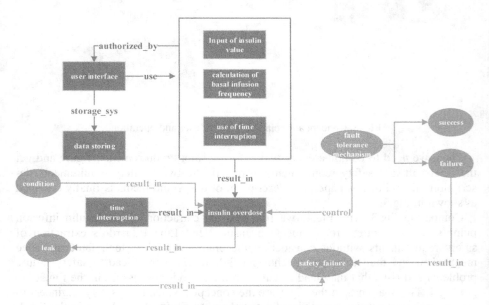

Fig. 5. The Model after supplement

4   Conclusion

In this paper, we use ontology to describe the safety requirements of the embedded system and set up a model. During this process, we define concepts that are the common safety requirements of the embedded system combined with its own characteristics. Then we adopt iterative confirmation to deduce the final safety requirements model. Our model can help safety analysts find the deficiency of their requirements and complete them. Our model is not focus on user's requirements, but also potential requirements (e.g. condition) which may appear in the process of software or systems development. This model also reduces user's extra operations during requirement analysis.

Comparing with other description methods of safety requirements, the advantage of constructing ontology model is:

- Describing articulated knowledge and tacit knowledge of safety requirements on the knowledge layer.
- Reflecting the relationships among objects within a certain domain on the logical layer.
- Defining concepts and terminologies within a certain domain explicitly and formally.

The analysis of ontology clarifies the structure of domain knowledge. Thus it can lay a good foundation for knowledge. Because ontology can be reused, it avoids repeatability of domain knowledge analysis. By constructing a unified framework or a standard model to reduce the differences in concepts and terms, making sharing and exchange of information possible among people who work in different fields or operating platforms.

In later work, we will focus on building a prototype program to support of the automation for the process and automatic validation of the model.

Acknowledments. This work was supported by the Jilin Province Science Foundation for Youths (20150520060JH).

References

1. Boehm, B.W.: Software Engineering. IEEE Transactions on Computers **25**(12), 1226–1241 (1976)
2. Firesmith, D.G.: Engineering safety requirements. J of Object Technology **2**(1), 53–68 (2003)
3. Kemmann, D.I.S., Trapp, M., Kalmar, D.I.R.: Safety analysis for embedded software. Atzelektronik Worldwide **4**(3), 10–15 (2009)
4. Amoroso, E.G.: Fundamentals of computer security technology. Prentice-Hall, Upper Saddle River (1994)

5. Haley, C.B., Laney, R., Moffett, J.D., et al.: Security Requirements Engineering: A Framework for Representation and Analysis. IEEE Transactions on Software Engineering **34**(1), 133–153 (2007)
6. Van Lamsweerde, A.: Elaborating security requirements by construction of intentional anti-models. In: Proceedings of the 26th Int'l Conf Software Eng(ICSE 04). IEEE CS Press (2004)
7. Leveson, N.G.: Software safety: why, what, and how. Computing Surveys **18**(2), 125–163 (1986)
8. Jin, Y.: Research on Eliciting Security Requirement Methods. Computer Science (2011)
9. Bo, L., Wei, L., Fei, W.: Modeling method of software security requirements based on UMLsec. Computer Engineering and Design **34**(9), 3124–3129 (2013). doi:10.3969/j.issn.1000-7024.2013.09.026
10. Chen, Z., Motet, G.: Formalizing safety requirements using controlling automata. In: Second International Conference on Dependability. DEPEND, pp. 81–86 (2009)
11. Kalloniatis, C., Mouratidis, H., Islam, S.: Evaluating cloud deployment scenarios based on safety and privacy requirements. Requirements Engineering **18**(4), 299–319 (2013)
12. Gruber, T.R.: A translation approach to portable ontology specifications. Knowledge Acquisition **5**(2), 199–220 (1993)
13. Xiao-Yong, D., Man, L., Shan, W.: A Survey on Ontology Learning Research. Journal of Software (2006)
14. Kaiya, H., Saeki, M.: Using domain ontology as domain knowledge for requirements elicitation. In: Proceedings of 14th IEEE International Requirements Engineering Conference, Minnesota, USA, pp. 186–195 (2006)
15. Hong-wei, W., Jia-chun, W., Fu, J.: A study on Ontology Model Based on Description logic. Systems Engineering **21**(3), 101–106 (2003)
16. Chung, L., do Prado Leite, J.C.S.: On non-functional requirements in software engineering. In: Borgida, A.T., Chaudhri, V.K., Giorgini, P., Yu, E.S. (eds.) Conceptual Modeling: Foundations and Applications. LNCS, vol. 5600, pp. 363–379. Springer, Heidelberg (2009)
17. Galvao Martins, L.E., De Oliveira, T.: A case study using a protocol to derive safety functional requirements from Fault Tree Analysis. In: 2014 IEEE 22nd International Requirements Engineering Conference (RE), pp. 412–419. IEEE (2014)
18. Zhang, Y., Jones, P.L., Jetley, R.: A hazard analysis for a generic insulin infusion pump. J. Diabetes Sci. Technol. **4**(2), 263–283 (2010)
19. Yaturu, S.: Insulin therapies: Current and future trends at dawn. World Journal of Diabetes **4**(1), 1–7 (2013)

Conceptual Framework for Understanding Security Requirements: A Preliminary Study on Stuxnet

Bong-Jae Kim[1](✉) and Seok-Won Lee[2]

[1] Department of Network Centric Warfare, Ajou University, Suwon, Republic of Korea
drzakal@ajou.ac.kr
[2] Department of Software Convergence Technology,
Ajou University, Suwon, Republic of Korea
leesw@ajou.ac.kr

Abstract. As we analyze the latest occurring Advanced Persistent Threats (APT), we understand that attackers tend to conduct their methods by various means and ways, and are not limited to just a single pattern. Even though threats are represented by using a variety of modeling techniques, very little research has been done to learn about the security requirements needed from the understanding of commonly rooted causes of these threats. We propose a layered conceptual framework to better understand the problem from specific instances to generalized abstractions, so that it can eventually provide a good set of security requirements. In this framework, we propose building the ontology based on the Goal-based Model and Activity Diagram, and showing how to understand the context of the problem domain with extended relationships between concepts. We expect that our work can provide a foundation of good insight of the problem domain of security requirements and that we elicit security requirements through this work based on a preliminary case study on Stuxnet.

Keywords: Security concept requirements framework · Security ontology · Security Goal-based model · Attack activity diagram

1 Introduction

Recently, many response reports that analyze a variety of threats and attacks are provided by information security companies and government. Unlike past years, recent attack patterns in reports involve various elements such as networks, software, physical elements and human activity to achieve its malicious goal of an Advanced Persistent Threat (APT). In addition, these threats exploit one or more Zero-Day Vulnerabilities as parts of its attack. [1] Since it is difficult to predict the means and ways of such threats, it is essential to develop a way to understand the nature of the threats with possible means to overcome the said threat.

Although various threat modeling approaches are proposed [2], there are some limitations to creating requirements from these models. The majority of modeling techniques just concentrate on technical analysis, while research about extracting, eliciting, and inferring common and generalized requirements is insufficient. Though

© Springer-Verlag Berlin Heidelberg 2015
L. Liu and M. Aoyama (Eds.): APRES 2015, CCIS 558, pp. 135–146, 2015.
DOI: 10.1007/978-3-662-48634-4_10

there is much work to provide techniques for attack modeling, for example, identifying various attack routes with visualization and calculating the most plausible attack route by using the stochastic method in each route, the conceptual framework to elicit common concepts has not been yet identified in this area. [3]

Thus, it is essential to research and study framework that can create requirements based on common and generalized concepts, because it still needs to progress under a complex security environment. In this paper, we have conducted research on understanding the problem domain using defined relationships between concepts and extended relationships based on the security conceptual framework, and creating security requirements. Using this framework, we expect to understand the problem domain of security requirements, generate models using reusable concepts, and help to draw the countermeasure using a requirements engineering process.

We organize four more sections to discuss our conceptual framework. Section 2 provides the introduction of related works for our research. Section 3 introduces the conceptual framework for our research, and Section 4 outlines the case study of our work using Stuxnet. Lastly, Section 5 concludes our research and provides the application to future work.

2   Related Work

We have conducted literature surveys to learn about relationships between security requirements and threats from the past researches.

First, a goal-model for the security area is suggested by Elahi et al [4] using i* framework, and they provide the comparison between i* framework and other conceptual framework in order to show the adaptability of i* framework to generate the security requirements as a goal model. In addition, they extended the i* framework to represent security notations such as vulnerability, malicious goals, etc., and adapted it into the case study, the Guardian Angel, to validate the goal model adaptability in the security area.

Lin et al. [5] used the modeling method of privacy and security based on the Guardian Angel case study, and suggested the concept of extraction based on the scenario, using a dependency analysis model for making connections between attack analysis models and countermeasure analysis models.

Research on the ontological engineering have been actively progressive and there are several efforts related to the security. First of all, Li et al [6] proposed the ontology based on the condition changes and behavior to design a system that provides the alert against intrusive behavior. However, it is hard to figure out the concept from the example, and to generate the corresponding security requirements from this model.

Wang et al. [7] proposed an ontology to manage the vulnerability related to the Common Vulnerability Enumeration (CVE) and Common Platform Enumeration (CPE), which give us some useful elements of security ontology. However, they did not make connections with real threats. Their works provide us the inspiration of a research base as we try to adopt the real threats.

Kotenko et al. [8] proposed the implementation of ontology in the SIEM system in order to overcome the limitations of the relational database, and their work give to us an understanding of the advantages of the implementation of ontology.

Elahi et al. [3] represented the conceptual framework with ontology as the meta-data model. They compared multiple conceptual frameworks including the i* frame-work, which can model malicious behavior and vulnerabilities with ontology. They considered malicious actions and countermeasures in their ontology. However, the context awareness of the various variant of tasks needed to be considered.

Lee et al. [9] proposed the Problem Domain Ontology Process to identify the security requirements for DITSCAP Automation, the Onto-ActRE and the modeling process using four steps of the PDO process. Their works defined the metrics and measures, and generated the security requirements. Through their works on the relationship between process, modeling and ontology, we understand that the conjunction between concepts in models and ontology provides the ability to generate the security requirements based on the surrounding knowledge.

Finally, we select the Stuxnet case as a case study of our approach which is a good example of Advanced Persistence Threats from [1]. The study of the Stuxnet is referred on [10, 11]. They analyze the malware with a technical approach and we referred to [12] for the modeling of the case.

3   Proposed Conceptual Framework

This section will provide imitations of the previous research efforts and the introduction to the proposed conceptual framework to overcome the limitations.

3.1   Approach

The generation of security requirements in information security is very challenging due to the complex and diverse attack techniques. To specify the requirements in the requirements engineering process, it is necessary to understand the problem domain.

The proposed framework consists of three layers: a physical layer, information layer, and cognitive layer. First, the physical layer represents real-world events, accidents, phenomena, and results of attacks, etc. The information layer is the area that consists of the analysis and representation of physical layer through the components of assets, attacks, countermeasures, goals, etc. Lastly, the cognitive layer is an understanding of both the physical and information layer. This layer can provide reasoning on the business, political, economic, society, standard, policies, and regulations impacts of threats and events in the physical and information layers.

In order to understand the security problem domain, we will propose the ontology based on the above three layers which can perform the bridge of understanding between each model and security requirement, and can create relationships between extracted concepts, like in Fig. 1. As you can see relations between layers in Fig. 1, artifacts in the information layer is generated through modeling and analyzing threats in the physical layer, and elements of artifacts in the information layer are mapped to generate the ontology based on the meta-data model. This ontology can provide the context-awareness specification using relations between concepts, and generate the scenario model. The scenario model will be the foundation for eliciting possible threats or real scenarios.

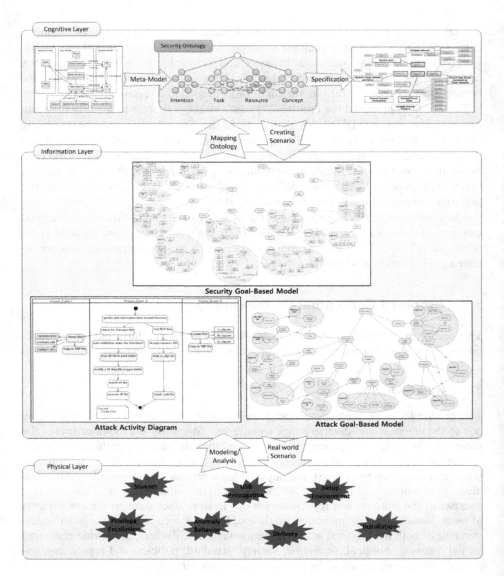

Fig. 1. The relationship between modeling and ontology based on layers

3.2   Overview of Security Concept Framework

In this sub-section, we briefly introduce our conceptual framework. As shown in Fig. 2, by comparing the Requirements Engineering Process with Stakeholder Requirements Definition Process in the System Security Engineering Process, published in the Special Publication 800-160 by the National Institute of Standard and Technology (NIST) [13], we will show how the framework can be implemented.

The Requirements Engineering Process consists of the following phases: Elicitation, Understanding, and Structuring, Modeling and Analysis, Communication and Negotiation, and Verification and Validation. This process is similar to the Stakeholder Requirements Definition Process in the System Security Engineering Process: Elicit Stakeholder Security Requirements, Define Stakeholder Security Requirements, and Analyze and Maintain Security Requirements. The conceptual framework performs part of this process and creates artifacts to understand the requirements during process. Ultimately, the purpose of the conceptual framework is to understand the related context of the given physical security situation, and to analyze the related security requirements components by considering the consequential results on laws, regulations. Then, one can understand the overall risks through the integrated ontology.

First, underlying this process, the fundamental research to identify instances related to concepts must be conducted. Next, based on these instances goals, tasks, and actors are extracted with an Attack Goal-Based Model (AGBM) using the i* framework, one of goal-based models in the requirements engineering. Then, an Attack Activity Diagram (AAD) is created, which can extract activities, related vulnerabilities, and resources based on identified task from the AGBM. Through mapping between elements in AGBM and AAD, a Security Goal-Based Model (SGBM) is generated, and instances related to security concepts in the SGBM are elements of the Security Ontology.

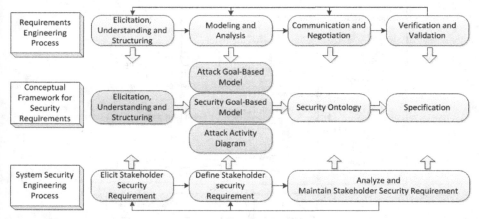

Fig. 2. Relationship with Requirements Engineering Process, System Security Engineering Process and Conceptual Framework for Security Requirements

Meta-data Model for Conceptual Framework

The Conceptual Framework is conducted based on the Meta-data Model in Fig. 3. These Security Concepts are elements in AGBM and ADD. Using this meta-model, we create the Goal-Based Security Model and finally the Security Ontology with the relationship among the domains.

First, we divide the domains into four categories: the Intention, Task, Concept, and Resource Domains. The Intention Domain includes both the Status and Goal knowledge with the hierarchical structure between them in the domain. The Goal is similar to the attack phase. We apply Bryant's work [14], which constructs the taxonomy of the attack intention. The Task Domain includes the Threat, Threat Version, Actor, and Task Instance with a vertical hierarchical structure. Threat means the name of the threat in Task Domain, and it has the version as the subclass. Moreover, the Malicious Actor is something that performs the task to achieve one or more malicious goals within each version. In addition, Task Instance means the real task conducted by the Malicious Actor. Concept Domain includes the conceptual knowledge related to the threat, such as the vulnerability and the related platform that can be implemented for the attack. In our paper, we use the Common Vulnerability Enumeration (CVE) and Common Platform Enumeration (CPE). The Resource Domain is related to the means or resources used to conduct the task: the Service and Policy Layer, Application and Software Layer, and Network Layer. The Service and Policy Layer mean the policy or activity performed by people or organizations. The Application and Software Layer is the same concept as the application layer in TCP/IP Protocol Stack, such as software, process, OS, etc. Lastly, the Network Layer is the abstracted layer from transport to the physical layer in the TCP/IP Protocol Stack.

These Security Concepts makes relationships with each other. The solid line in the diagram means the vertically hierarchical connection. The arrow line indicates the relationship between different domains or different areas. These defined relationships are basis of the inference and assumption.

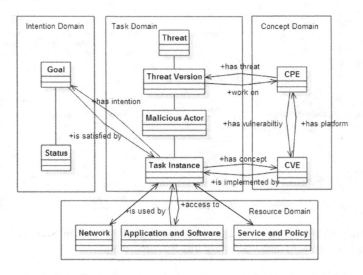

Fig. 3. Security Concept Requirement Framework Meat-Data Model

Each concept in this meta-data model came from elements of models in the information layer. Through eliciting elements during the framework, each element forms the ontology with relations between concepts in order to make aware the contexts of threats. For example, elements in the Intention and Task Domains came from AGBM which used the i* framework, and elements in Concept and Resource Domain came from AAD.

3.3   Attack Goal-Based Modeling

Previous works for attack modeling techniques were proposed in various forms, such as Attack Graph, Attack Tree, Bayesian Graph with a stochastic approach, and the Boolean Drive Markov Process (BDMP) models [2]. In the case of these models, they provided the modeling technique for specific events. However, as they focus on modeling with a technical analysis, there are limitations in the perspective of Requirements Engineering, such as limited representation of the intention and the trade-off value and reinterpretation for implementing the Requirements Engineering Process. To remedy these drawbacks, we propose the Attack Goal-Based modeling (AGBM) rooted from Goal-based modeling in the Requirements Engineering Process, which represent the intention and trade-off value.

Reasons why we choose the Goal-based model out of various requirements engineering models include the fact that people who have malicious intentions conduct almost all threats in cyber space, and these intentions are the foundation of tasks of customized malicious actor or techniques. To achieve the malicious goal, malicious people build certain methods and tools that can access resources to perform tasks and a series of activities. Though these elements of attack are implemented into codes and programs, we can extract concepts through a series of abstracting, modeling, and analysis of various attack scenarios to generate security requirements not only for these scenarios, but also for those that can be prevented early because generated requirements consider the common characteristics of various attack/threats scenarios.

Attack Goal-Based Modeling (AGBM) uses i* Framework with the OpenOME, the open source tool for goal-based modeling. Elahi et al. [4] conducted the research on frameworks comparison for the security requirement representation with other frameworks. They showed an explicit representation of relationships with goals, tasks, soft-goals, and resources and the extensibility for the security notation. Therefore, we choose i* framework because of strong points when illustrating the chained-attack using the embedded and extended notation. In this model, tasks will be related to the Attack Activity Diagram, and resources and related vulnerabilities will be shown.

3.4   Attack Activity Diagram

An Attack Activity Diagram (AAD) is the modeling process used to extract resources and concepts related to each task with serialized order of activities, and identify the anomaly behavior and interaction with resources. There are three reasons why we choose the Activity Diagram in UML for modeling the detail part. First, because we notice that the chained attack is conducted across multiple layers and has certain steps to perform the task, it is necessary to illustrate temporal-ordered events and behaviors that are difficult to represent in a goal model effectively. In addition, it needs to represent the multi-layered concept for understanding layered resource access. Lastly, the activity diagram gives the specification of the behavior process.

3.5   Security Goal Based Model

Security Goal-Based Model (SGBM) is the result of the integration with mapping instances in the AAD into the AGBM in order to understand the threat flow. Like the AGBM, it also uses the i* Framework and extends the previous work with embedded notations, resources and concepts from the AAD. Activities in AAD are represented as the concept of context-awareness with the interaction of resource. Therefore, the SGBM looks like the extended version of AGBM with adding resources and concepts. Because SGBM represents the specific event, this model needs to be converted into the ontology. In other words, SGBM constructs a bridge between Security Ontology, real events, and foundation of the ontology.

3.6   Security Ontology

This is integrated ontology based on the meta-data model extracting instances related to security concepts from previous models. Through using the ontology, we can define the relationship between security concepts. Advantages of the security ontology are that it provides a more effective comparison with the relational database model, in which it is difficult to change the scheme, gives the context awareness based on accessing resources and relationships with other information, and provides the inferences to understand other concepts from the specific concept. In addition, the specification of the problem domain can be means for the communication among related stakeholders, and be helpful in forming the common understanding for generating requirements. Protégé 4.3[15] based on the Meta-Data Model is the editor application for this ontology.

4   Case Study

This section provides the case study of the proposed conceptual framework, which shows the process from modeling to creating the ontology using the Stuxnet case in 2010. First, Stuxnet has the purpose of sabotaging the Iranian Nuclear Program in the Natanz Uranium Enrichment Plant, by exploiting the vulnerability of the Industrial Control Systems (ICS), Programmable Logic Controller (PLC), and Siemens Step-7 software with Windows systems that operate in the isolated network. They deliver the malware from an external environment using the thumb drive to plug into the isolated network, automatically propagate, and conduct the attack against inside and outer network. It also spreads worldwide with high infection probability via the Internet, more specifically located in the Middle East. It used several techniques, such as Zero-day vulnerability Exploitation, Windows Rootkit, PLC Rootkit, Antivirus Evasion Technique, Network infection routines, Command and Control Interface, etc. [11, 12] In this case study, we will briefly show the case of   Export 15.

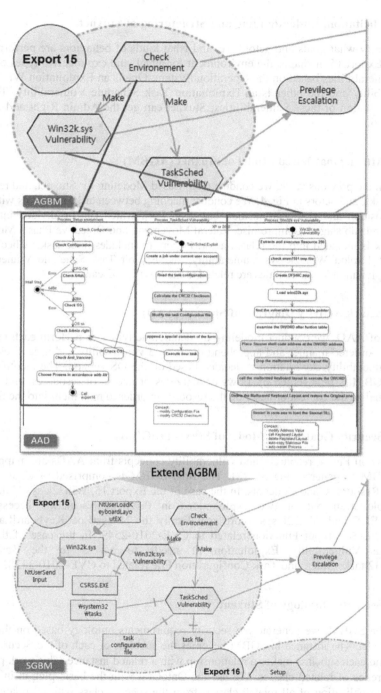

Fig. 4. How to make relationships between AGBM, AAD, and SGBM

4.1   Elicitation, Understanding and Structuring of Stuxnet

We identify what goals and actors are, and what kinds of behaviors are performed. In case of Export 15, it shapes the environment for attack and exploits two types of Zero-day Vulnerabilities based on the Operation System. One is an Exploitation Win32k.sys Vulnerability and the other is an Exploitation Task Schedule Vulnerability. Through the exploitation of these vulnerabilities, Stuxnet can get the Admin Right and call the Export 16.

4.2   Attack Goal-Based Model of Stuxnet (AGBM)

Based on the previous work, we conduct Goal Based Modeling for Stuxnet, and represent goals, tasks, and actors in Fig. 4. We conduct mapping between tasks and goals with And-decomposition and Or-decomposition based on [14], which can be divided into three phases and sub stages: Compromise, Lateral Movement, and Objective Phase. Moreover, each task is included into the related actor. Export 15 includes three tasks: Check Environment, Exploit Win32k sys Vulnerability, and Exploit Task Schedule Vulnerability. Two Exploitation Vulnerabilities are related to the Privilege Escalation.

4.3   Attack Activity Diagram of Stuxnet (AAD)

In case of AAD, we conduct modeling on detailed activities to perform each task in a temporal-ordered manner and apprehend the related vulnerability and access resources. Each vulnerability exploitation is decided based on the OS version. The identified task from AGBM is represented by flows of activities or access of resources, and we can conceptualize the context or describe the resource in order to map them into the SGBM.

4.4   Security Goal-Based Model of Stuxnet (SGBM)

As shown in Fig. 4, resources and vulnerability concepts from AADs are mapped into the AGBM to generate the SGBM. The final SGBM is comprised of Goals, Tasks, Actors, Resources, and Concepts. In the case of the Export 15, Win32k.sys Vulnerability Exploitation out of two exploitations in Privilege Escalation accesses the CSRSS.EXE and Win32k.sys, more specifically the NtUserLoadKeyboardLayoutEx and NtUserSendInput Function related to CVE-2010-2549. In the case of the Task Schedule Vulnerability Exploitation, it needs to access to the Task file, \system32\tasksfolder, and Task Configuration File related to CVE-2010-3338.

4.5   Security Ontology of Stuxent

Using the SGBM, we generate the security requirements ontology based on the Meta-Data Model. The instance in SGBM becomes the subclass of each of the Security Concepts, and each subclass makes a relationship with a related instance. This work provides the inference of related classes. We use the OntoGraf with the basic plug-in in Protege 4.3 for visualization of all related classes from the specific class with all relationships, and this is a strong point of comparison with other visualization plug-ins.

In Fig. 5, we show one example of the visualization about the Exploit Task Schedule Vulnerability and related Security Concepts. First, this task is used to get the Admin Right as the Privilege Escalation. Then, we get the specification including information in previous section: for example, the threat name is Stuxnet, version information is following the detected date, 20100719, the Actor is Export 15, the related CVE is CVE-2010-3338, and the related platform is Windows Vista and the Server 2008 Series.

Though this example just shows only one case, it will become very powerful after gathering knowledge from many cases. As we provided the case study, we understand inferred ideas from the specific concept through the ontology. If we extend the ontology with the aid of knowledge base, it provides various understandings of the context of the situation. Moreover, extending the ontology with the countermeasure, we expect to provide the countermeasure priority and the risk assessment.

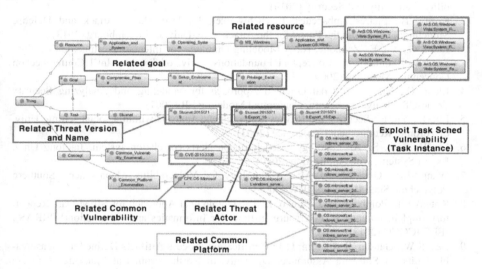

Fig. 5. The visualization for related classes of 'Exploit Task Schedule Vulnerability' using OntoGraf

5   Conclusion and Future Work

Until now, we have discussed the conceptual framework, implemented into the case study using Stuxnet, and have shown how it works. Through this process, we represent the specifications and visualizations of security concepts, and propose the ontology based on the goal-based model. We expect that our research provides understanding of the problem domain in the security requirements in perspective of the requirements engineering process, and generating requirements.

However, as this is the first step of our preliminary research, we need to take further steps, such as extending the ontology including countermeasure and quality attributes, and conducting the research on the detection based on similarities against variant threats and the risk assessment using the security ontology in order to consolidate our framework. Ultimately, our final goal is to develop a solid framework based on these kinds of work in order to generate security requirements.

Acknowledgments. This research was supported by the Basic Science Research Program through the National Research Foundation of Korea (NRF) funded by the Ministry of Education, Science and Technology (No. 2013R1A1A2009801).

References

1. Virvilis, N., Gritzalis, D.: The Big Four- What we did wrong in Advanced Persistent Threat Detection?. Athens Univ. of Economics and Business, Int'l Conference on Availability, Reliability and Security (2013)
2. Kordy, B., Pietre-Cambacedes, L., Schweitzer, P.: DAG-Based Attack and Defense Modeling: Don't Miss the Forest for the Attack Tree. Univ. of Luxembourg (2013)
3. Elahi, G., Yu, E., Zannone, N.: A Modeling Ontology for Integrating Vulnerabilities into Security Requirements Conceptual Foundations. Univ. of Toronto, Int'l Conference on Conceptual Modeling (2009)
4. Elahi, G., Yu, E.: A Goal Oriented Approach for Modeling and Analyzing Security Trade-Offs. Univ. of Toronto, Conceptual Modeling ER (2007)
5. Lin, L., Yu, E.: Security and Privacy Requirement Analysis within a social setting. Univ. of Toronto, Requirement Conference (2003)
6. Li, W., Tian, S.: An Ontology-based intrusion alerts correlation system, Jiaotong Univ. Expert Systems with Application (2010)
7. Wang, J.A., Guo, M.: OVM: An Ontology for Vulnerability Management. Southern Polytechnic State Univ., CSIIRW (2009)
8. Kotenko, I., Polubelova, O., Senko, I.: The Ontological Approach for SIEM Data Repository Implementation. St. Petersburg Instituet for Informatics and Automation(SPIIRAS), IEEE ICGCC (2012)
9. Lee, S.W., Gandhi, R.A., Ahn, G.J.: Certification Process Artifacts Defined as Measureable Units for Software Assurance. The Univ. of North Carolina at Charlotte, Software Process Improvement and Practice (2007)
10. Falliere, N., Murchu, L.O., Chien, E.: W32.Stuxnet Dossier, Symantec, Security Response (2011)
11. Matrosov, A., Rodionov, E., Harley, D., Malcho, J.: Stuxnet Under the Microscope, ESET (2010)
12. Kriaa, S., Bouissou, M., Pietre-Cambacedes, L.: Modeling the Stuxnet Attack with BDMP: Towards More Formal Risk Assessments
13. Ross, R., Oren, J.C., McEvilley, M.: System Security Engineering: An Integrated Approach to Building Trustworthy Resilient Systems. National Institute of Standards and Technology, NIST Special publication 800–160 (2014)
14. Bryant, B.: A Method for Implementing Intention-Based Attack Ontologies with SIEM Software. Fishnet Security, Securely Enabling Business White Paper (2014)
15. Horridge, M.: A Practical Guide To Building OWL Ontologies Using Protégé 4 and CO-ODE Tools Edition 1.3, the University Of Manchester (2011)

Requirements Engineering Tools

TimePF: A Tool for Modeling and Verifying Timing Requirements Based on Problem Frames

Yuanyang Wang, Xiaohong Chen[✉], and Ling Yin

Shanghai Key Laboratory of Trustworthy Computing,
East China Normal University, Shanghai, China
xhchen@sei.ecnu.edu.cn

Abstract. As the key element of embedded systems, the timing requirements are becoming more and more important. An increasing number of systems need strict time constraints especially some safety critical systems such as high-speed railway systems. We have proposed an approach for modeling and verifying timing requirements [1–3]. By combining the Problem Frames (PF) approach [4] and CCSL (Clock Constraint Specification Language) [5], it can model timing requirements from the perspective of environment and verify them with NuSMV. To support this approach, we develop a supporting tool (named TimePF) by extending the DPTool [6]. TimePF is a graphical tool which provides various modeling and verifying techniques for timing requirements. This paper presents its architecture and implementation, and gives an illustrating example to show how to use it following the modeling and verifying process.

1 Architecture and Implementation

1.1 Architecture

Fig.1 shows the architecture of TimePF. It includes 3 layers, i.e., data layer, function layer and interface layer. The function layer models and verifies timing requirements. The data layer provides data for function layer. And the interface layer is responsible for the interactions with users. We will introduce the function layer in detail. It includes modeling module and verifying module. The modeling module includes 2 sub-modules, i.e., data processing module and graphical operation module.

Modeling Module
- Data processing sub-module
This module accepts the inputs in terms of a problem diagram and scenario graphs. All the problem domains in problem diagram will be traversed to find interactions related to each problem domain. Meanwhile, all the paths in scenario diagram will be traversed to find the possible temporal relations among interactions.
- Graphical operation sub-module
This module edits graphic elements on the interface. It adds a clock for each interaction. All the clocks and clock relations form a clock diagram. In addition, this module defines composite clocks and additional constraints. These new added constraints will be checked to ensure that there are no conflicts among the new constraints and existing ones.

© Springer-Verlag Berlin Heidelberg 2015
L. Liu and M. Aoyama (Eds.): APRES 2015, CCIS 558, pp. 149–154, 2015.
DOI: 10.1007/978-3-662-48634-4_11

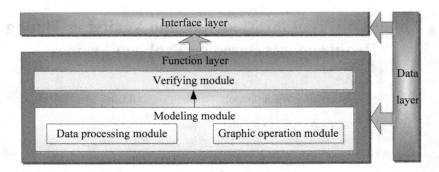

Fig. 1. Architecture of TimePF

Verifying Module

The verifying module transforms the CCSL description from the model into NuSMV description, and verifies the consistency of timing specification. The transformation follows the transforming rules. It also defines the consistency properties with CTL, and verifies these properties by calling NuSMV.

1.2   Implementation

The TimePF is implemented in Java. Fig.2 shows its snapshot. There are 5 major areas on the interface, i.e., *Menu*, *Toolbar*, *Process*, *Plotting* and *Information area*.

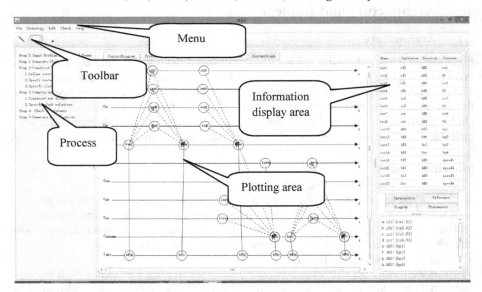

Fig. 2. Snapshot of TimePF

- There are 5 items in the *Menu*, i.e., *File, Ontology, Edit, Check, Help*. The *File* menu is responsible for creating, opening and saving a project. The *Ontology* menu can load, show and check environment ontology. The *Edit* menu edits operations about clock and clock constraint. And the *Check* menu executes some checking operations such as *Check consistency*. The *Help* menu shows some help information about the tool.
- The *Toolbar* gives three kinds of qualitative relations among clocks. They are *precedence, coincidence* and *strict precedence*.
- The area of *Process* shows the steps to be followed in the tool.
- The *Plotting area* shows the diagrams drew by the tool, including problem diagram, scenario diagram, clock diagram and time point diagram.
- The *Information area* shows some information related to the diagram in the *Plotting area*. The information includes diagrams, phenomenon, interactions and citation information.

2   Process and Illustrating Example

Fig.3 shows the process of modeling and verifying timing requirements. It includes 2 major steps, i.e., modeling timing requirements, and verifying timing specification. In order to be clearly understood, we use the Anti-lock Braking System (ABS) as an example to show the steps of using TimePF. The description of ABS is as follows.
It is composed by four sensors and four actuators. The four sensors (ifl, ifr, irl, irr) are used to measure the rotational speed of wheels. And the four actuators (ofl, ofr, orr, orl) represent the braking pressure on each wheel. The ABS is triggered by R. The signals of four sensors must arrive in a certain input delay (for example 0.5ms).

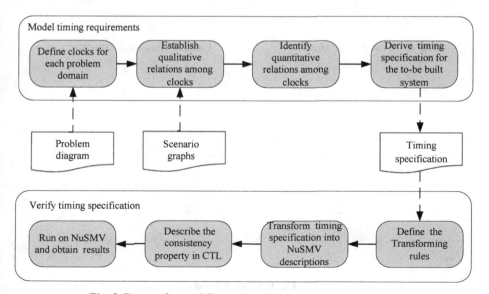

Fig. 3. Process for modeling and verifying timing requirements

Step 1: Model Timing Requirements. The inputs of this step are a problem diagram and scenario graphs. The output is the timing specification of the to-be built system. It has 4 sub-steps:

1) Define clocks for each problem domain in the problem diagram. A clock is defined as $C:=< I, <>$, where, C is the clock, I is the set of time points, and < is the partial order relation defined on I which named *StricPre*. Clocks are classified into two kinds, domain clocks and interaction clocks. The domain clock is expressed as $d.C$, where d represents the problem domain and C represents the clock. And the interaction clock can be defined as *int.C,* where *int* represents the interaction and C represents the clock. Since each problem domain can initiate or receive many interactions, each domain clock is composed by the interaction clocks of this domain using the union operator of CCSL [6]. If a domain is composite, its clock will be established by the clocks of sub-domains using clock operators of CCSL including *sup, inf* and *union* [6].

Fig. 4. Define domain clocks

Fig. 5. Define clock constraints

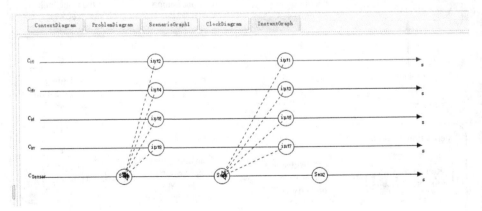

Fig. 6. Clock relations

Example: In the ABS system, int_1 is an interaction that ABS software initiates and domain ifl receives R. It has a corresponding clock, $int_1.C_1$. Domain ifl has a clock too, $ifl.C_{ifl}$. As domain ifl has two related interactions, int_1 and int_2, then the Time-PF automatically generates $ifl.C_{ifl} = int_1.C_{1ifl}$ union $int_2.C_{2ifl}$. In the tool, Clock editor (Fig. 4) defines domain clock, and union constraints are shown in Fig. 6.

2) Establish qualitative relations among clocks. The qualitative relation operators provided by CCSL consist of *subclock, fasterThan* and *alternate*. These qualitative relations could be obtained directly from existing scenario graphs. We argue that interaction relations of the same problem domain could be directly transformed to the following clock relations:

- *int_1 behOrd int_2* implies C_{int1} strictPre C_{int2}.
- *int_1 behEna int_2* implies C_{int1} strictPre C_{int2}.
- *int_1 reqOrd int_2* implies C_{int1} strictPre C_{int2}.
- *int_1 syncBehReq int_2* implies C_{int1} subClock C_{int2}, and C_{int2} subclock C_{int1}.

Example: In the scenario graph of ABS, int_1 behOrd int_2, then we get C_{int1} strictPre C_{int2} (Fig. 6). This is generated automatically.

3) Identify quantitative relations among clocks expressed as C_1 *bounderDrift(i, j)* C_2. It means that the time interval between the time points C_1 and C_2 is within the range $[i, j]$, where i is a negative integer and j is a positive integer.

Example: In the ABS, int1 happens before int2 is within the range [1, 2], then we use bounderDrift relation as shown in Fig.5.

4) Derive the timing specification of the to-be built system. Firstly, define clock C_{sys} for system by *union* each problem domain clock. The timing specification contains three parts: system clock C_{sys}, related clocks of C_{sys} and clock relations including the qualitative relations obtained from step 2) and quantitative relations obtained from step 3).

Example: In the ABS system, we define C_{ABS} for the system. It is composed by Sensor and Actuator. So we get $C_{ABS} = C_{Sensor}$ union $C_{Actuator}$. The Sensor clock has relations with other clocks as shown in Fig.6. Finally, a specification in format of .txt is obtained.

Table 1. Transformation rules from timing specification to the NuSMV descriptions

Timing specification in LTS	NuSMV descriptions
Clock C	C:boolean
Clock relation	constraint MODULE
State s	VAR s:boolean;
Transition s {C_1, C_2,...,C_n} s'	TRANS case s=TRUE: next(C_1)=TRUE & next(C_2)=TRUE &...& next(C_n)=TRUE & next(s)=FALSE & next(s')=TRUE case;

Step 2: Verify Timing Specification. The input of this step is the timing specification obtained in step 1. And the output of this step is the verification results and the verified timing specification. It has 4 sub-steps:

1) Define the transformation algorithm from the timing specification to the descriptions of NuSMV. Firstly, we define the operational semantics of timing specification using Labelled Transition System (LTS). By direct mapping, the transformation rules form LTS to NuSMV descriptions are given in Table 1. Then the transformation algorithm could be defined (omitted due to limited space).

2) Transform the timing specification into the NuSMV descriptions according to the transformation rules in 1).

Example: *According to the transformation algorithm, the constraint ifl.C_{ifl}= int$_1$.C_{1ifl} union int$_2$.C_{2ifl} could be transformed into ctr1: union (C_1, C_2, C_{ifl}). According to the transformation rules in Table 1, union could be transformed to:*

> MODULE **union**(left, right, new)
> TRANS
>   (next(left)=TRUE & next(new)=TRUE)|(next(right)=TRUE &
> next(new)=TRUE)|(next(left)=FALSE & next(right)=FALSE & next(new)=FALSE)

3) Describe the consistency property of timing specification in CTL. For the timing specification with clock set T= $\{C_1, C_2,...,C_n\}$, we say it is consistent if it satisfies the following two CTL formula:

  o $EF(AGp)$, where $p=!(C_1|C_2|...|C_n)$;
  o $\forall C_i \in T$, $EF\ (AGq)$, where $q=!C_i$.

Example: *In the ABS system, $T=\{C_{ABS}, C_{Sensor}, C_{Actuator}, C_{ifl}, C_{ifr}, C_{irl}, C_{irr}, C_{ofl}, C_{ofr}, C_{orl}, C_{orr}, C_{1ifl}, C_{2ifl}, C_{1irr}, C_{2irr}, C_{1irl}, C_{2ifr}, C_{2irl}, C_{1ifr}, C_{3ofl}, C_{4ofl}, C_{3ofr}, C_{4ofr}, C_{3orr}, C_{4orr}, C_{3orl}, C_{4orl}, C_{int1}, C_{int2}, C_{int22}, C_{int3}, C_{1Sensor}, C_{1sensor2}, C_1, C_2, C_3, C_4, C_5, C_6, C_7, C_8, C_9, C_{10}, C_{11}, C_{sen1}, C_{sen2}, C_{sen3}, C_{act1}, C_{act2}, C_{act3}\}$. Its consistency formula is generated automatically by our tool.*

4) Check the consistency of timing specification using NuSMV, and obtain the verification results and the verified timing specification.

Example: *After clicking Check consistency menu, the consistency of the ABS system could be checked automatically. There is no inconsistency condition. Then the timing specification could be output in the format of .txt at the same folder with the project.*

Acknowledgement. This research was supported by the National Natural Science Foundation of China under grants 61202104 and 91418203, and the Doctoral Fund of Ministry of Education of China under Grant 20120076120016.

References

1. Chen, X., Liu, J., Mallet, F., Jin, Z.: Modeling timing requirements in problem frames using CCSL. In: Proceedings of the 18th Asia-Pacific Software Engineering Conference (APSEC 2011), pp. 381–388 (2011)
2. Yin, L., Chen, X., Liu, J.: Consistency analysis of timing requirements for cyber-physical system. Journal of Software, 25(2), 400−418 (2014, in Chinese)
3. Chen, X., Liu, J.: Modeling Software Timing Requirements: An Environment Based Approach, 36(1), 88–103 (2013, in Chinese)
4. Jackson, M.: Problem Frames: Analyzing and Structuring Software Development Problems. Addison-Wesley (2001)
5. Mallet, F.: Clock constraint specification language: specifying clock constraints with UML/MARTE. Innovations in Systems and Software Engineering. **4**(3), 309–314 (2008)
6. Chen, X., Yin, B., Jin, Z.: DPTool: a tool for guiding the problem description and the problem projection. In: Proceeding of the 18th IEEE International Requirements Engineering Conference, pp. 401–402 (2010)

Using an Ideas Creation System to Assist and Inspire Creativity in Requirements Engineering

Delin Jing[1](✉), Chi Zhang[1,2], and Hongji Yang[1]

[1] Centre for Creative Computing, Bath Spa University, Corsham SN13 0BZ, UK
{delin.jing13,chi.zhang14,h.yang}@bathspa.ac.uk,
lytzhangchi@buu.edu.cn
[2] Tourism Institute, Beijing Union University, Beijing 100101, China

Abstract. Software product markets have become extremely competitive as there are always multiple software products striving to serve the users in the same application domain. In order to be successful, a software system needs to distinguish itself from other similar products and surprise users with novel and useful features. Obviously, creativity becomes much more important in a software engineering process, especially for requirements, as creative requirements engineering is crucial to new and surprising features or services. However, normally, with it focuses on elicitation, analysis, and management, research studies on requirements engineering do not offer strong support to creativity in requirements engineering. Naturally, services like idea generation can be involved to support creativity requirements engineering by eliciting innovative ideas from stakeholders. Although a considerable number of applications and research studies have been made in the past years in order to increase the effectiveness of idea making process, there is little work exists to design an ideas creation system for assisting and inspiring requirements. Meanwhile, it is lack of efforts working on creativity requirements in the requirements engineering perspective particularly. Therefore, the objective of this research paper is to propose an ideas creation system to assist engineering activities for generating creativity requirements. In particular, this paper designed an ideas creation framework and defined and classified a set of creativity elements according to creativity techniques. Then, it proposes a creative requirements engineering method that is supported by the designed ideas creation system and creativity elements, whilst the application domain is specific to the e-learning service. An inference engine is the kernel part in the idea generation process with domain ontology for the target field as the knowledge base. Hence, the generated ideas are inspiring stakeholders to get not only relevant and useful but also novel and surprising requirements.

Keywords: Requirements engineering · Creative requirements · Ideas creation · Creative computing · Creativity

1 Introduction

Traditionally, Requirements Engineering (RE) considers that requirements exist in the stakeholders' minds in an implicit manner [1], and focuses on models and techniques to aid identification and documentation of such requirements [2]. Current software product market, however, has become extremely competitive as there are always

© Springer-Verlag Berlin Heidelberg 2015
L. Liu and M. Aoyama (Eds.): APRES 2015, CCIS 558, pp. 155–169, 2015.
DOI: 10.1007/978-3-662-48634-4_12

normally multiple software products striving to serve the users in the same application domain [2]. In order to sustain and be successful, a software system needs to surprise customers with novel and useful features [2]. Therefore, creativity is necessary to be involved to achieve this target, especially for requirements elicitation, because creative requirements are the beginnings of new and surprising features or services. However, existing studies to requirements engineering offers not much support to creativity. Idea generation, as a way to inspire individual or team members to generate more and new ideas, can be used as the fundamental process of getting innovative outcomes in various domains. Therefore, this paper suggests that a proper designed ideas creation system can be involved to support creativity in requirements by providing new, useful and surprising requirements to inspire and elicit innovative and clear requirements from stakeholders.

This research aims to provide an ideas creation system to assist engineering activities for creativity in requirements. In the following sections, firstly, background knowledge is explained following by reviews of related work. Secondly, an ideas creation framework is proposed with explanations of different phases. Next, based on creativity techniques, creativity elements are defined and classified, whilst corresponding rules of creativity elements' application are designed. The ideas creation system is to provide ideas as information that is able to help stakeholders to get clear and innovative requirements. Supported by creative computing techniques, including exploration, combination and transformation, the generated ideas are inspiring stakeholders to get not only relevant and useful but also novelty and surprising requirements. After the requirements elicited, they are presented as a mind map with special tags corresponding to defined creativity elements to indicate the requirements' various and specific demands on creativities, which makes the requirements formally formatted and provides convenience for the subsequent application design and development. Last, a case study is presented to illustrate how the proposed method works. Overall, the main contribution of this paper is the designed ideas creation system.

2   Background and Related Work

2.1   Requirements Engineering and Creative Requirements

Along with arise of creativity in software engineering, requirements engineering community has received a growing interest from researchers and practitioners. It emerges many papers discussed on the high level of creativity and requirements such as work from Lemos and his colleagues [1] and discussions from Maiden [3, 4] and Bhowmik [2]. Beside, some research studies worked on providing various techniques to do requirements engineering in creative ways, such as using Model-Driven Engineering [5], mind mapping [6] and reasoning [7]. Although there are efforts, they are not mature methods yet and it is still lack of ability to be implemented. Therefore, as mentioned in last section, this paper concentrates on using an ideas creation system to support creativity in requirements engineering. In particular, the ideas creation system is designed to provide clues to assist and inspire stakeholders on eliciting clear and innovative requirements. Moreover, a set of creativity elements are designed to help on evaluating generated ideas in creativity perspective and to support requirements presentation by combining with mind mapping technique.

2.2   Creativity and Creative Computing

Creativity is an extremely important facet of life and is a feature of many of the tasks that people do every day. It can occur in a multitude of situations ranging from work to pleasure, from artistic portrayals to technological innovation [8]. Most texts regard creativity as a beneficial process in an organisation and it has been said to offer a competitive advantage in the design processes [9]. Naturally, it is a crucial feature for new and innovative ideas; consequently, creativity needs to be considered in the idea generation process.

According to Boden's definition [10], an idea can be called "new" from two perspectives: the objective (H-creative) and the subjective (P-creative) view. They derive from two kind of creativity: H-creativity (short for historical creativity) and P-creativity (short for psychological creativity) [11, 12]. H-creativity is fundamentally novel in respect to the whole of human history and P-creativity is the personal kind of creativity that is novel in respect to the individual mind [13, 14]. From the above discussion, the creativity expected in this paper should belong to H-creativity. Obviously, H-creative ideas are very difficult to be generated by individual or a group, especially on the creativity perspective, because it requires to be supported by extensive knowledge and creative techniques.

Because creativity is considered the ultimate human activity and a highly complex process [15], some researchers hold that the creative thinking process cannot be formulated, analysed, or reconstructed [13], [15, 16]. Others adopt a reductionist view that creative products are the outcome of ordinary thinking, only quantitatively different from everyday thinking [10], [14], [17]. By review related studies and developments, this research believes that creative ideas can be generated systematically, if there is a carefully designed supporting method, which is an ideas creation system in this research.

Similar to the definition of creativity, there is not a universal definition of creative computing. In last few years, creative computing is being discussed more widely, hoping to produce new, innovative and valuable products. Creative computing seeks to reconcile the objective precision of computer systems (mathesis) with the subjective ambiguity of human creativity (aethesis) [11]. As a newly aroused emerging research field, in creative computing, there are many promising research directions have been studied [18, 19], such as creative design, creative requirement engineering, and creative collaboration. One research objective in creative computing is to find the approach to get creativity and to realise it [11], [20]. Besides, creative computing can be recognised as the study of computer science and related technologies and how they are applied to support creativity, take part in creative processes, and solve creativity related problems. Creative application software, or called as creative application can be referred to those software, tools, or environment which can support, improve or enhance creativity using text, graphics, audio, video, and integrated technologies [21, 22]. Currently, there are researches working on approach and process to develop creative software from the beginning. However, it does not exist ideas creation system for creativity in requirements engineering.

3   An Ideas Creation Framework

In an earlier published paper [23], we proposed an ideas creation process as Figure 1 shows, which is a high level process suitable for general idea generation purposes. Specifically, there are three kernel phases to create new ideas including "Knowledge Extraction/Reuse", "Idea Generation" and "Ideas Evolution". Based on this process, this paper is to design a specific ideas creation system for assisting and inspiring creativity in requirements engineering. Thus, it focuses on the ontology construction and ideas' creativity evaluation, which belong to the first phase "Knowledge Extraction/Reuse" and the last phase "Ideas Evolution". A set of creativity elements is defined in next section to support the ideas' creativity evaluation. Besides, combined with mind mapping technique, the creativity elements supports requirements presentation. The following contexts explain more details on the three phases in Figure 1 to illustrate the entire ideas creation process.

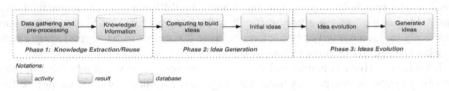

Fig. 1. Ideas Creation Process

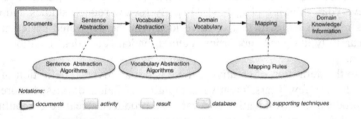

Fig. 2. Knowledge Extraction/Reuse in the Ideas Creation Process

Phase 1: Knowledge Extraction/Reuse. As Figure 2 shows, this phase is data gathering and pre-processing by adopting abstraction techniques, designed abstraction algorithms and mapping rules [24], and reusing knowledge bases [25]. The extraction part works as following description. Firstly, it determines the objective of task and selects relevant documents as raw data. Then the domain vocabulary is extracted from the text data supported by abstraction algorithms. Last, the extracted domain vocabulary is mapped into the ontology format to be the domain knowledge/information according to designed mapping rules. Because building an ontology from scratch is not only time consuming but also limited to gathered resources, moreover, the ontology based domain knowledge is reusable, thus, it is more efficient to reuse existing domain ontologies to assist the construction of specific domain knowledge base. In particular, there are two circumstances in the knowledge reuse: 1) if a knowledge ontology exists for the required domain but is not up to date, it requires a smaller scale knowledge extraction to get the latest information and then merges the extracted information into the existing domain ontology to form the requisite knowledge base; and 2) if there is a knowledge ontology extracted recently for the required domain, the existing domain ontology will be reused directly as the knowledge base for the subsequent idea generation. In this paper, because the ideas creation system is designed to support creativity in requirements engineering, the knowledge base is ontology of requirements, which is more focusing on functions, features, etc. Furthermore, since the application field is narrowed down to e-learning service, the ontology is constructed for e-learning service. The above two points distinguish the ideas creation system from others.

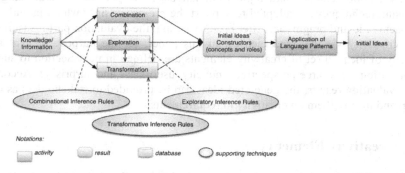

Fig. 3. Creative Idea Generation in the Ideas Creation Process

Phase 2: Idea Generation. It is computing to build ideas as Figure 3 shows. Based on gathered and processed knowledge, system computes following designed algorithms and rules to generate initial ideas to realise convergent thinking. Exploration, transformation and combination are kernel activities of its computing step to generate initial ideas. Combination activity involves unfamiliar combinations of familiar knowledge and information. Exploration activity explores within an established conceptual space. This is more likely to arise from a thorough and persistent search of a well-understood space. Transformation activity deliberately transforms a conceptual space. It should involve the rejection of some of the constraints that define this space and some of the assumptions that define the problem itself. These three kinds of activities provide the basis of the techniques to compute resources and generate initial ideas. The results of one activity can be input of another activity to generate ideas through multi-activities. However, it is not necessary to implement all three kinds of activities. The practical realities of their application must be worked out in different applications and circumstances, usually on a case-by-case basis. A set of inference rules has been designed in our previous publication [26] to support the proposed three activities as supporting techniques. Moreover, language patterns are proposed [26] to be applied to form the generated ideas as readable phases and sentences.

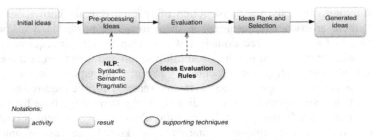

Fig. 4. Ideas Evolution in the Ideas Creation Process

Phase 3: Ideas evolution. It covers pre-processing, evaluation, ranking and selection as Figure 4 shows. The pre-processing is to analysis the generated initial research ideas as a good sentence and topic from language perspective. Natural Language Processing techniques are adopted to support the pre-processing, which include syntactic, semantic and pragmatic. After pre-processing, the evaluation part measures the ideas' creativity via designed metrics. In this paper, for the purpose of evaluating creativity of ideas, a set of creativity elements are defined in next Section to support the evaluation from three perspectives: novelty, usefulness and surprising. According to the evaluation results, the generated ideas can be provided to stakeholders as clues to help and inspire them on creativity in requirements.

4   Creativity Elements

Dean et al. [27] carried out an exhaustive analysis of studies that employed criteria to assess creativity in solution and product ideas [27, 28]. Based on review and analysis of selected 51 relevant studies, and for purposes of their quantitative tool, these researchers further broke these criteria into more specific and measurable terms including dimensions and sub-dimensions as shown on Table 1. In order to conquer the inconsistencies limitation on the previous studies, Dean and his colleagues [27] recommended adopting the naming conventions depicted in Table 1. Their research discussed that it helps to avoid confusion between novelty-only studies and creativity studies where creativity is based on novelty plus other quality constructs [27]. However, a creative outcome is unlikely can be effective if it is novelty only.

Besides, according to the definitions, novel, workable, relevant, and specific are on the same level while each of them has a set of sub-dimensions in the lower level. In the relationships refers to Figure 5, workable, relevant and specific are in the same level while novel is in a higher level. In hierarchical perspective, the relationships conflicts with the definitions of the constructs.

Fig. 5. Relationships among Constructs/Dimensions [27]

Table 1. Definitions of Quality Dimensions and Sub-dimensions [27]

#	Dimension	Definition
1	*Novelty*	The degree to which an idea is original and modifies a paradigm.
1.1	Originality	The degree to which the idea is not only rare but is also ingenious, imaginative or surprising.
1.2	Paradigm relatedness	The degree to which an idea is paradigm preserving or paradigm modifying.
2	*Workability (Feasibility)*	An idea is workable (feasibility) if it can be easily implemented and does not violate known constraints.
2.1	Acceptability	The degree to which the idea is socially, legally, or politically acceptable.
2.2	Implementability	The degree to which the idea can be easily implemented.
3	*Relevance*	The idea applies to the stated problem and will be effective at solving problem.
3.1	Applicability	The degree to which the idea clearly applies to the stated problem.
3.2	Effectiveness	The degree to which the idea will solve the problem.
4	*Specificity*	An idea is specific if it is clear (worked out in detail).
4.1	Implicational explicitness	The degree to which there is a clear relationship between the recommended action and the expected outcome.
4.2	Completeness	The number of independent subcomponents into which the idea can be decomposed, and the breadth of coverage with regard to who, what, where, when, why, and how.
4.3	Clarity	The degree to which the idea is clearly communicated with regard to grammar and word usage.

Overall, in our opinion, the proposed constructs, sub-dimensions and the relationships cannot directly employed in this research. There are limitations and conflicts in various levels. However, some of these researchers' methods are worth to be adopted. Specifically, this research proposes a set of creativity elements and corresponding sub-dimensions by adopting Dean and his colleagues' [27] way to define constructs. Also, similarly, a hierarchical structure is useful on the relationships among the proposed creativity elements and sub-dimensions.

Boden [10] says a creative idea is novel, surprising, and valuable. Most important, creative ideas should be surprising because they go against out expectations [10]. That is to say, a creative idea should be not only rare but also be ingenious and imaginative. Thus, this research proposes three creativity elements: Novelty, Usefulness and Surprising. Novelty measures the idea is new from different perspectives. Usefulness is to make sure an idea is applicable and is worthy of study, which covers valuable, but not only that. Comparing with Dean and his colleagues' constructs, Usefulness actually contains relevance, workable and specific but with improvement to overcome their limitations and conflicts. Surprising measures the degree of ideas' unexpectedness and unusualness, that is how much ideas against out expectations and how much unique the ideas are, which distinguishes Surprising with Novelty.

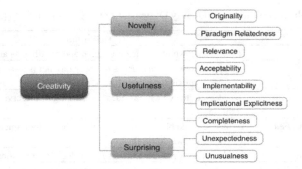

Fig. 6. Relationships of Creativity Elements and Sub-Elements

This research adopted some of sub-constructs from Dean et al. [27], redefined them accordingly, and re-catalogued into the proposed elements as their sub-elements. Figure 6 shows the hierarchical relationships of the creativity elements and sub-elements. In the following context, when the creativity elements and sub-elements are considered as a whole, it is referred as Creativity Elements (CE) for convenience.

4.1 Definitions of Creativity Elements

For the purposes of this research, which is to generate ideas for creative requirements of e-learning service, the proposed creativity elements and sub-elements are defined or redefined, depends on if it exists in previous research studies and suitable to be adopted. The following provide a more in-depth explanation of the respective elements.

- **Novelty**: The degree to which an idea is original and modifies a paradigm of e-learning requirements.
  - *Originality*: The degree to which an idea is rare in H-creativity perspective. The low degree means the idea is common, mundane, boring. The high degree means the idea is not expressed before.
  - *Paradigm relatedness*: The degree to which an idea is paradigm preserving (PP) or paradigm modifying (PM). PP ideas remain same concepts or same relationship between concepts with a paradigm. PM ideas extend concepts, or redesign or transform relationship between concepts. PM ideas are sometimes radical or transformational.
- **Usefulness**: An idea is useful if it can be easily implemented as a requirement and does not violate known constraints in the domain knowledge. It should be relevant to the specific domain or domains as well as workable (feasible) as a requirement.
  - *Relevance*: The idea applies to research in specific domain/domains and will be effective as a research according to user's input. In other words, it covers both domain relevance and input relevance.
  - *Acceptability*: The degree to which the idea is acceptable (not conflict knowledge constraints). Low acceptability means the idea violates knowledge constraints. High acceptability means the idea does not violate knowledge constraints.

— *Implementability*: The degree to which the idea can be easily implemented as a requirement. Low implementability means the idea is hard to achieve or hard to get valuable outcomes as a function or feature. High implementability means the idea can be implemented as a requirement well.

— *Implicational explicitness*: The degree to which there is a clear relationship between the recommended action and the expected outcome. Low implication explicitness means the implication in the idea is not stated or less relevant. High implication explicitness means the implication in the idea is clearly stated and makes sense.

— *Completeness*: The number of independent subcomponents into which the idea can be decomposed, and the degree of the subcomponents expressed in the idea.

- **Surprising**: It is about the unexpected degree of the ideas.

  — *Unexpectedness*: The degree of the idea goes against out the user's expectation.

  — *Unusualness*: The degree of the idea distinctiveness, that is how much unique the idea is.

4.2   Creativity Elements for Requirements Engineering

The above creativity elements are proposed to support requirements engineering, particularly on requirements presentation. To achieve this aim, it is designed to combine with the designed ideas creation system and mind mapping technique to support requirements elicitation and presentation respectively as Figure 7 shows. In particular, the creativity elements work in the ideas evolution phase to classify the generated ideas according to the defined creativity elements. If a generated idea is classified into one or more creativity elements and adopted as a requirement by the stakeholders, corresponding tags for creativity elements will be added into the specific requirement's node when the requirements are presented as a mind map. Thus, the majority affect and aim of the creativity elements is to bring clear vision of creativity on requirements presentation.

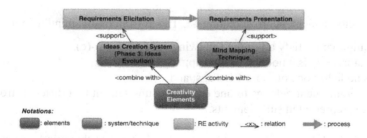

Fig. 7. Creativity Elements for Requirements Engineering

4.3   Designed Tags and Rules for Creativity Elements

This paper adopts tags as the way to mark creativity in requirements. Each creativity element and sub-element has a corresponding tag, which is designed as <element name> format and named as "CE Tag". Table 2 lists all the creativity elements with their corresponding CE Tags. The CE Tags indicate the requirements' various and specific demands on creativities.

Table 2. Creativity Elements and Corresponding CE Tags

Creativity Elements	CE Tags
Novelty	<novelty>
Originality	<originality>
Paradigm relatedness	<paradigm relatedness>
Usefulness	<usefulness>
Relevance	<relevance >
Implementability	<implementability >
Acceptability	<acceptability >
Implicational Explicitness	<implicational explicitness >
Completeness	<completeness >
Surprising	<surprising>.
Unexpectedness	<unexpectedness >
Unusualness	<unusualness>

Besides, mind mapping is employed as a basic technique to support requirements presentation. Following rules are designed for applying the above designed CE Tags into the mind mapping process. Basically, the designed rules can be classified into two categories as below shows,

Rules for Creativity Sub-elements:

If a requirement satisfy all of the following conditions (a)-(c),

(a) a requirement is a node in mind map;

(b) this node has no CE Tag; and

(c) this requirement belongs to creativity sub-elements.

then

add this sub-element's corresponding CE Tag in front of this mind map node.   (1)

If a requirement satisfy all of the following conditions (a)-(c),

(a) a requirement is a node in mind map;

(b) this node has one or more CE Tags; and

(c) this requirement belongs to one creativity sub-element that different from existing CE Tags represented sub-elements.

then

add this sub-element's corresponding CE Tag in front of this mind map node; and merge it with other CE Tags in this node.   (2)

Rules for Creativity Elements:

If a requirement satisfy all of the following conditions (a)-(d),

(a) a requirement is a node in mind map;

(b) this node has at least one sub-node;

(c) this node has no CE Tag; and

(d) there are sub-elements' CE Tags in one or more of its sub-nodes.

then

add CE Tag in this node and the added CE Tag represents creativity element that contains the sub-elements corresponding to the CE Tags in condition (d).   (3)

If a requirement satisfy all of the following conditions (a)-(d),

(a) a requirement is a node in mind map;

(b) this node has at least one sub-node;

(c) this node has at least one CE Tag; and

(d) there are sub-elements' corresponding CE Tags in one or more of its sub-nodes that not included in this node's CE Tag.

then

add CE Tag in this node, while the added CE Tag represents creative element that contains the sub-elements corresponding to the CE Tags in condition (d).   (4)

If a requirement satisfy all of the following conditions (a)-(d),

(a) a requirement is a node in mind map;

(b) this node has at least one CE Tag;

(c) this node has no father-node between itself and root node; namely, this node is directly linked to root node; and

(d) it cannot be categorised into another node.

then

add a father node for this node, give the father node an abstract name, and add high level creativity elements' CE Tags according to CE Tags in condition (b).   (5)

If a requirement satisfy all of the following conditions (a)-(d),

(a) a requirement is a node in mind map;

(b) this node has at least one CE Tag representing creativity sub-element;

(c) this node has no father-node between itself and root node; namely, this node is directly linked to root node; and

(d) it can be categorised into another node.

then

link this node with the other node that is identified in condition (d) as a sub-node; and run rule (3) or (4) for its new father node depending on whether this father node's has CE Tag.   (6)

5   Case Study

This section discusses requirements engineering for a Chinese (Mandarin) e-learning application to demonstrate and prove that the proposed ideas creation system is feasible to be applied to support creativity in requirements engineering. Its ultimate goal is to provide an ingenious application allows users to explore innovative ways to learn Chinese. An ontology of e-learning service is the first thing needed for the ideas creation system. As there is not exists an ontology suitable to be used directly, a new ontology of e-learning service is constructed as the knowledge base. Figure 8 shows part of the ontology's graphical representation in Protégé.

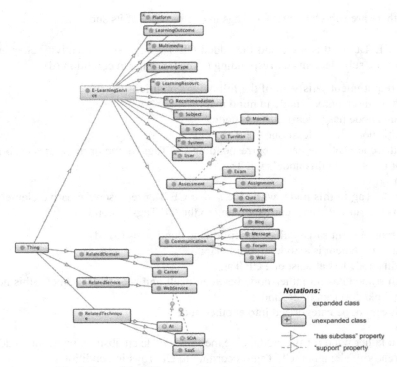

Fig. 8. Part of the E-Learning Service's Ontology in Protégé

Details about idea generation phase are omitted here as it is followed the designed approach and rules in our earlier papers [25], [29]. Figure 9 shows the input and output screenshots of the developed ideas creation system. A user can be a stakeholder, an application designer or a domain expert. To use the ideas creation system, a user needs to select a domain and enter some keywords as the system input. The domain is the application's domain, whilst keywords are other relevant information. Concerning this case study, it is suitable to select "Language E-Learning" as the domain and to enter "Chinese", "Mandarin" and "Application" as keywords. Through the idea generation process, the final generated results are listed in the output interface.

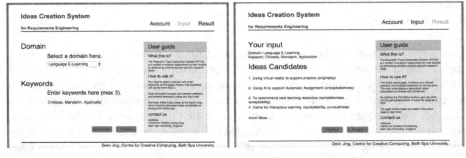

Fig. 9. Input and Output Screenshots of Ideas Creation System

The defined creativity elements help to evaluate creativities for generated ideas as requirements. After adopting all the generated four ideas, the requirements are presented in a mind map with CE Tags as Figure 10 shows, which is ready to be processed in the further software engineering phases to develop the target application.

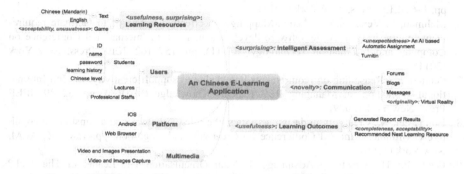

Fig. 10. Final Requirements Mind Map for the Chinese E-Learning Application

6   Conclusion

This research paper proposed an ideas creation system to assist engineering activities for creative requirements development, which aims to cope with the software domain's rapid development and intense competition. In particular, an ideas creation framework is designed with three phases to provide ideas as clues to help and inspire requirements elicitation. Also, creativity elements are defined and classified according to creativity techniques, which are to evaluate creativities in generated ideas and to support requirements acquisition via proposed CE tags and designed rules. The applied CE Tags indicate the requirements' various and specific demands on creativities, which make the requirements formally formatted and provide convenience for the subsequent design and development. Furthermore, a Chinese e-learning application is selected as a case study to illustrate and prove the feasibility of the proposed ideas creation system for creativity in requirements engineering.

References

1. Lemos, J., Alves, C., Duboc, L., Rodrigues, G.N.: A systematic mapping study on creativity in requirements engineering. In: 27th Annual ACM Symposium on Applied Computing (SAC 2012), pp. 1083–1088. ACM, New York (2012)
2. Bhowmik, T., Niu, N., Mahmoud, A., Savolainen, J.: Automated support for combinational creativity in requirements engineering. In: 22nd IEEE International Requirements Engineering Conference, pp. 243–252. IEEE Press, New York (2014)
3. Maiden, N.: Creativity in software engineering: a new research Agenda? In: 18th IEEE International Conference on Program Comprehension, pp. xiv. IEEE Press, New York (2010)

 4. Maiden, N., Ncube, C., Robertson, S.: Can requirements be creative? Experiences with an enhanced air space management system. In: 29th International Conference on Software Engineering, pp. 632–641. IEEE Press, New York (2007)
 5. Assar, S.: Model driven requirements engineering: mapping the field and beyond. In: 4th IEEE International Model-Driven Requirements Engineering Workshop (MoDRE), pp. 1–6. IEEE Press, New York (2014)
 6. Mahmud, I., Veneziano, V.: Mind-Mapping: An effective technique to facilitate requirements engineering in Agile software development. In: 14th International Conference on Computer and Information Technology (ICCIT), pp. 157–162. IEEE Press, New York (2011)
 7. Schmid, K.: Reasoning on requirements knowledge to support creativity. In: 2nd International Workshop on Managing Requirements Knowledge (MARK), pp. 32–39. IEEE Press, New York (2009)
 8. Bonnardel, N.: Creativity in design activities: the role of analogies in a constrained cognitive environment. In: 3rd Conference on Creativity and Cognition, pp. 158–165. ACM, New York (1999)
 9. Cook, P.: The Creativity Advantage: Is Your Organisation The Leader of The Pack? Industrial and Commercial Training, vol. 30, no. 5, pp. 179–184. MCB University Press, Bradford (1998)
10. Boden, M.A.: The Creative Mind: Myths and Mechanisms, 2nd edn. Routledge, London (2004)
11. Yang, H., Hugill, A.: The Creative Turn: New Challenges for Computing. International Journal of Creative Computing 1(1), 4–19 (2013). Inderscience
12. Hugill, A.: Creative computing processes: musical composition. In: 8th IEEE International Symposium on Service Oriented System Engineering, pp. 459–464. IEEE Press, New York (2014)
13. Koestler, A.: The Act of Creation. Hutchinson, London (1964)
14. Perkins, D.N.: The Mind's Best Work. Harvard University Press, Cambridge (1981)
15. Goldenberg, J., Mazursky, D., Solomon, S.: Creative Sparks. Science 285, 1495–1496 (1999)
16. Bono, E.D.: Lateral Thinking: Creativity Step by Step. Harper & Row, New York (1970)
17. Zhang, J., Ma, J., Zhang, D., Tan, R., Source, A.K.: CAI-Driven new ideas generation for product conceptual design. In: IEEE International Conference on Management of Innovation and Technology (ICMIT), pp. 824–830. IEEE Press, New York (2012)
18. Carroll, E.A., Latulipe, C., Fung, R., Terry, M.: creativity factor evaluation: towards a standardised survey metric for creativity support. In: 7th ACM Conference on Creativity and Cognition, pp. 127–136. ACM Press, New York (2009)
19. Nguyen, L., Shanks, G.: A Framework for Understanding Creativity in Requirements Engineering. Information and Software Technology, vol. 51, pp. 655–662 (2009). Butterworth Heinemann, Newton, MA
20. Jing, D., Yang, H., Xu, L., Ma, F.: Developing a creative idea generation system for innovative software reliability research. recently published In: 2nd International Conference on Trustworthy Systems and their Applications. (2015)
21. Janssen, D., Schlegel, T., Wissen, M., Ziegler, J.: MetaCharts-using creativity methods in a CSCW environment. In: Human-Computer Interaction Theory and Practice (Part II), pp. 939–943. Mahwah, New Jersey (2003)
22. Weiley, V., Pisan, Y.: The distributed studio: towards a theory of virtual place for creative collaboration. In: 20th Australasian Conference on Computer-Human Interaction: Designing for Habitus and Habitat, pp. 343–346. ACM Press, New York (2008)

23. Jing, D., Yang, H.: Domain-Specific 'Idea-tion': Real Possibility or Just Another Utopia? Recently Published in: Applied Science Journal (2015)

24. Jing, D., Yang, H., Tian, Y.: Abstraction Based domain ontology extraction for idea creation. In: 13th International Conference on Quality Software (QSIC), pp. 341–348. IEEE Press, New York (2013)

25. Jing, D., Yang, H., Shi, M., Zhu, W.: Developing a research ideas creation system through reusing knowledge bases for ontology construction. In: 39th IEEE Annual Computers, Software and Applications Conference (COMPSAC), pp. 175–180. IEEE Press, New York (2015)

26. Jing, D., Yang, H.: Creativity techniques based inference activities and rules for idea generation. In: 3rd International Symposium on Software Technology, pp. 1–8. (2015)

27. Dean, D.L., Hender, J.M., Rodgers, T.L., Santanen, E.L.: Identifying Quality, Novel, and Creative Ideas: Constructs and Scales for Idea Evaluation. Journal of the Association for Information Systems 7, 646–698 (2006)

28. Puccio, G.J., Cabra, J.F.: Idea generation and idea evaluation: cognitive skills and deliberate practices. In: Mumford, M.D. (ed.) Handbook of Organisational Creativity, pp. 189–215. Elsevier Inc, London (2012)

29. Jing, D., Yang, H.: Creative computing for bespoke ideation. In: 39th IEEE Annual Computers, Software and Applications Conference, pp. 34–43. IEEE Press, New York (2015)

EVALUATOR: An Automated Tool for Service Selection

Muneera Bano[✉] and Didar Zowghi

Faculty of Engineering and Information Technology,
University of Technology Sydney, Ultimo, NSW, Australia
{Muneera.Bano,Didar.Zowghi}@uts.edu.au

Abstract. The large number of third party services creates a paradox of choice and make service selection challenging for business analysts. The enormous online reviews and feedback by the past users provide a great opportunity to gauge their sentiments towards a particular product or service. The benefits of sentiment analysis have not been fully utilized in third party service selection. In this paper we present a tool that assists the business analysts in making better decisions for service selection by providing qualitative as well as quantitative data regarding the sentiments of the past users of the service. The tool has been applied and evaluated in an observational case study for service selection. The results show that sentiment analysis helps in increasing relevant information for business analysts, assists in making more informed decisions, and allows us to overcome some of the challenges of service selection.

Keywords: Service selection · Sentiment analysis · Requirements engineering

1   Introduction

Although Service Orientation was proposed as a new style of software development to addresses some of the shortcomings of previous approaches [1], it has inherited some of the challenges of component based and object oriented development, in particular in the requirements engineering [2, 3]. In Service Oriented Requirements Engineering (SORE) an analyst has an additional challenging task of aligning requirements and services to select the optimally matched service from an increasingly large set of available online services [4, 5]. Due to large number of online services offering similar functionality, the analysts require additional source of information for making more informed decisions for service selection [6].

User involvement in software development has been the focus of significant research and has been intuitively and axiomatically accepted to play a positive role in users' satisfaction thus leading to system success [7, 8]. More recently, past users' feedback, reviews and comments from online sources have been considered a form of user involvement [9-11]. These offer valuable information to assist analysts in increasing their knowledge for making more informed decision for service selection [12]. The user comments and feedback have been major sources of evolution of Android market and Apple store applications [13-15]. Online user feedback and sentiment analysis has attracted great interest in various areas of software engineering

© Springer-Verlag Berlin Heidelberg 2015
L. Liu and M. Aoyama (Eds.): APRES 2015, CCIS 558, pp. 170–184, 2015.
DOI: 10.1007/978-3-662-48634-4_13

research e.g. Requirements Elicitation [15, 16], Software Evolution [11, 17], and Software Quality [13]. 'Sentiment Analysis' (also known as opinion mining) is used for calculating and monitoring the attitude and behaviour of the past users from their feedback, comments and reviews available on the online social media. Various Sentiment Analysis tools, techniques and methods [11], are proposed that make use of Natural Language Processing, Computational Linguistics, Text mining and analytics capabilities for calculating quantitative values of various users' attitude and behaviour towards a particular product [10]. In service oriented paradigm the full extent of the benefits of this form of user involvement has not been empirically investigated [18].

In this paper, we present a tool 'EVALUATOR' that supports our previously proposed ARISE (Alignment of RequIrement and SErives) method [18, 19]. In ARISE method, we have explored the benefits of past user feedback analysis on the process of service selection and have evaluated its usefulness for analysts in overcoming the challenges of alignment. The tool aims to assist the business analysts in making better decisions for service selection by providing qualitative as well as quantitative data regarding the sentiments of the past users of the service. We have applied EVALUATOR to the data collected from an observational case study [12], to assess the utility and working of the tool. EVALUATOR automates some aspects of ARISE in order to reduce the time and effort required for implementation of the method. The results show that sentiment analysis helps in increasing information for business analysts, assists in making better informed decisions, and overcoming challenges of service selection.

2   Background

2.1   Challenges of Service Selection

Identification of the correct service is the most important step in Service Oriented Software Engineering (SOSE) [20, 21]. According to the qualitative study involving interviews with practitioners, selecting a service against customers' requirements is considered a challenging task due to the following reasons [2, 22]:

- Services are developed free of context to cater the needs of large number of customers. The lack of contextual information in service description or specification makes it challenging to decide about the suitability of the service in a particular system.
- The advertisements of the third party online services published by the service providers often provide incomplete or ambiguous information.
- The functionality offered by the services is usually not at the same level of granularity as the customers' requirements.
- The level of abstraction in description of service specifications and customers' requirements are usually not at the same level.
- Due to availability of huge number of online third party services with similar functionality and cost, it is a paradox of choice when it comes to selecting the best match service for customers' requirements.

The existing solutions for service selection are focusing more towards the technical aspect of the challenges of service selection and the social aspects are neglected to the larger extent [2, 21].

2.2   User Involvement and System Success

It has been axiomatically accepted in the existing literature of four decades that user involvement in software development leads to successful systems [8, 23, 24]. The form of involvement basically describes the way in which the users are involved. There are three levels of user involvement [25]: *Informative, Consultative and Participative.* In consultative and informative roles, the users are required to provide the necessary information that can impact the decision making processes of the system development, and their physical presence is not necessary. In service oriented paradigm, user involvement is needed in order to provide systems that can be customized for individual user needs [9]. Past users of the service may not be known and available at the time of service based design and development but there is significant amount of feedback, reviews and comments available of individual services on social media, forums and blogs by the previous users of the service. For service oriented development, the past users of the service can be approached through their 'voice' from online resources and their feedback can be analyzed to elicit the require information.

2.3   Sentiment Analysis

In recent years, there has been a substantial body of research for proposing methods, tools and techniques on collecting and analyzing past users' feedback that is available online, comments and review for extracting useful information [11, 13-17, 26-28] (e.g. data mining, information retrieval, crowd sourcing, parsing, sentiment analysis). The user comments and feedback have been major sources of evolution in product line release in case of mobile apps. In service oriented domain, past user feedback can serve these purposes:

- Providing the information about the previous users' satisfaction based on their past experience of using the service. This will also reflect users' trust of service provider (if the service is from third party). While analysing past users' satisfaction it is important to consider the context in which the previous users have used the service. User feedback without context may not be useful at all.
- User feedback can be used for filling the gaps in service specification where the information is missing against the checklist that is developed in previous steps. Service specification may not be at the same level of abstraction as customer requirements in giving details about functional and non-functional capabilities of service. User feedback can help in identification of missing information in service specification [17].
- Overcome challenges of alignment in SORE; The past users' feedback and sentiment analysis can help the analysts in overcoming the challenges of alignment process by:

— Collecting contextual data based on the previous usage of the service
— Finding and retrieving the missing information in service specification
— Comparing the service specification details against the real use of the service to bring the requirements and service specification on the same level of abstraction
— Eliciting past users' satisfaction level with the service and the reputation of service providers
— Monitor the popularity of a particular service among the users in case of multiple similar choices
— Making better informed decisions for service selection

This provided the motivation for proposing a method for Alignment of RequIrements and SErvices (ARISE) which utilises past users' feedback, sentiments and comments in the decision making process for the service selection. Next section presents brief summary of ARISE method. Full details of the method are presented in [18, 19].

3   ARISE – Service Selection with User Feedback

ARISE method [18] takes input of customer requirements, available service specifications, and past user feedback. The analyst uses ARISE to find the optimally aligned service among available options that best fits the customer preferences. The optimally aligned service here is defined as the *"one that satisfies maximum set of customer requirements (both functional and non-functional) according to their preferences while at the same time has good reputation with the past users"*. The ARISE method involves four different actors in the process of alignment:

- **Customers:** are the project sponsors for whom the service oriented software system is being developed, who have supplied the requirements, and will actually use this system in future. As with any software development projects, the customers in the project participate in various activities like requirements elicitation, modification, and prioritisation based on their preferences
- **Service Providers:** offers the services and advertises specifications or descriptions for the services they provide by publishing them either online or in the relevant organizational repository
- **(Past) Users:** are those who have experience of using a particular service in the past and have either provided feedback on online resources or can provide (post deployment) feedback when requested. This group can include the analysts, developers, designers who have previous experiences of actually using a particular service in software development. Their feedback is either collected form online resources (if available) or elicited directly from the users (if approachable).
- **Analysts:** are those who perform requirements elicitation, requirements prioritization, service searching, service specifications analysis and making decisions for service selection in current project at hand by following the steps of the ARISE method.

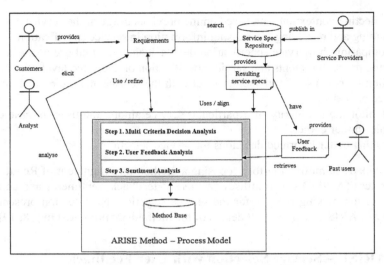

Fig. 1. Process Model for ARISE [18, 19]

Figure 1 represent the process model of ARISE method. The process of alignment in ARISE method starts with the elicitation of an initial set of requirements from customers represented by R such that $R = \{R1, R2, R3 \dots RX\}$ where X is the total number of requirements. Using the requirement set the analyst would search for available related services from accessible service repositories (local or global). Resulting services from this search are represented by $S = \{S1, S2, S3 \dots SY\}$ where Y is number of services found against requirement set R. The analysis in the ARISE method comprises of three interconnected and iterative steps [18, 19]: Multi Criteria Decision Analysis, User Feedback Analysis, and Sentiment Analysis. A "method base" is required for providing suitable tools, techniques, or methods for these three steps according to project situation and context. This provides the flexibility to ARISE method to be adopted for various project situations.

3.1   Multi Criteria Decision Analysis

The first step requires the analysts to evaluate all service specifications for granularity level against requirements and score them for their level of granularity using Multi Criteria Decision Analysis (MCDA) [29]. MCDA is used for decision making in situations where a trade-off is required among multiple criteria. The proposed method ARISE makes use of MCDA for scoring and ranking of services during alignment process. This step aims to score and rank the service set S in order to select the service that provides maximum functional range against requirements R i.e. a service that provides more coverage of requirement set. This step helps in filtering a sample of most relevant services from the set of available services which is manageable for further analysis. During this step the analyst converts the requirements into a checklist and assigns the weights to the checks based on customer preferences. The set of checks is represented by $C = \{C_1, C_2, C_3 \dots C_K\}$ and the weights against these checks is represented by $W = \{W_1, W_2, W_3 \dots W_K\}$ where K is the number of checks in the list. These weights provide prioritisation of the requirements based on customer preferences as not all the requirements are equally important for the customers. Various MCDA methods are available for different situations which can be stored in the Method Base. The most commonly used method of MCDA is Additive Weights method which in its simplest form assigns weights as multipliers to their respective checks or criteria (based on customer preferences) and then all scores for one option are added. The service with highest score is considered to be possibly best aligned among available options based on customer preferences. The assumption is that quantifiable weights are to be provided in the same unit of measurement for scoring by the customer based on their prioritization of the requirements. If not possible, then Aspiration level Methods [29] are available where the preferences are considered in their natural way rather than converting them all into one scoring level. A more dynamic approach is Outranking Method [29] which takes a more dynamic perspective and constructs preferences based on the information of available decision alternative rather than creating them before the actual analysis and decision making. While scoring services, there can be three main scenarios of alignment for a specific requirement from set R: fully aligned, totally misaligned, or partially aligned. The scores can be calculated by evaluating a service in one of these three scenarios: fully aligned (score 1), totally misaligned (score 0), or partially aligned (score between 0 to 1). For a service Si from the set of services S the score is represented by Score(Si) which is calculated by adding all the answers to the K number of checks in set C for that service according to the following formula.

$$\text{Score } (S_i) = \sum_{i=1}^{K} ((C_i) * (W_i))$$

Once the scores are calculated, the analyst can filter the sample highest scoring services that are most relevant to the requirements according to customer preferences. This will reduce the over burden of further analysis.

3.2   User Feedback Analysis

User feedback can be collected:

1. Directly from online sources if the users are unknown and not approachable
2. Elicited directly from the known and approachable users.

While aligning the services against requirements, there is a possibility that some information (especially performance related) might be missing in service specification. The analyst is required to assess the missing information (for its type and context) because it will be extracted from the feedback of previous users of the service. If new information is found then the analyst can go back to step 1 and update the MCDA scores for that specific service evaluation. There are various methods (and associated tools and techniques (e.g. feature extraction, information retrieval, crowd sourcing, survey and questionnaire etc.)) available for feedback collection based on the situational factors related to the availability of past users and the format in which the feedback is available (blogs, forums, twitter etc.). The user feedback can help the analysts in alignment process with

- Finding the missing information in service specifications or descriptions as advertised by the service providers, this would further help in
  - *Matching level of granularity of requirements and services by increasing the knowledge about service specifications or descriptions*
  - *Matching level of abstraction of requirements and services by increasing the knowledge about service specifications or descriptions*
- Collecting contextual data based on the previous use of the service
- Eliciting past users' satisfaction level with the service and the reputation of service providers
- Comparing the service specification details against the real use of the service

3.3   Sentiment Analysis

The process makes use of existing methods and tool in the fields of natural language processing, text analysis and computational linguistics in order to identify and retrieve required information from the sources. This provides quantifiable scores for ranking and comparison of product from different suppliers by using the online user comments and feedback and ratings as the source. Various sentiment analysis approaches and associated techniques and tools are available for gauging the reputation of a service by monitoring the sentiments of the users regarding that service. The selection of any specific depends on the situation regarding the type of input and output information required by the analyst.

Once the MCDA scores and sentiments scores are available, a comparison can be made among the services. For all Y number of services, the highest service score among the set S that has high sentiment scores as well, would be considered optimally aligned service according to the customer preferences.

4   Case Study

For instantiation of the ARISE method, we have previously conducted a case study. The preliminary results of the manual implementation of ARISE method in a case study were presented in [12]. It was perceived to be an appropriate methodology due to the following reasons; *(1) To observe the working of the method on a project, and to refine and improve ARISE method by applying the steps using data from real world; (2) To validate the idea of involving user feedback in the service selection process for overcoming challenges of alignment in SORE; (3) To find the requirements for an automated tool support for ARISE method.*

The case study was observational in nature. This case study presents the practical implementation of the ARISE method in a real world project and the effect of involving user feedback in overcoming the challenges of service selection process are identified. The hypothesis that guided the design of the case study is: *"User feedback assists in overcoming challenges of aligning requirements and services"*. The case selected for the evaluation of the ARISE method involved the selection of a Short Message Service (SMS) gateway service for an existing website. SMS gateway services enable websites to send and receive text or multimedia messages from a web browser over telecommunication network to a mobile device with simple invocation of the remote service API while hiding all the underlying technical and infrastructure details. These services provide a ubiquitous and seamless way to the developers to approach roaming users allowing communication capabilities from the websites. The gateway services also act as a translator from one network protocol to another and connect different SMS centres that handle various operations such as receiving, storing, or forwarding SMS to the desired destination mobile network. These services have given a great opportunity for the companies to stay connected to their clients via their mobile devices. The website in this case belongs to a gym that required the SMS facility in their online system to contact its registered members considering that it is faster to approach them through SMS rather than email. There are currently 500 registered members. All of the members are located in Sydney. The gym needs to send single, group, or broadcast messages only in Australia to its members for different notification purposes such as registration expiry, new offers, change in timings etc. The frequency of sending one SMS to a member of gym is higher than sending bulk messages to all the members. Most of the services provide cheaper rates when SMS are sent in bulk, Therefore the gym is looking for a service that provides cheaper rates per one SMS as well. In addition to the cost and the basic functional requirements, the reliability and timely communication are the top priorities especially when sending activation codes to mobile phones for online registration of new members.

Table 1 shows some of the 28 checks that were created based on customer requirements that were used for evaluation of the services [12]. Manual online searches (the searches were conducted in June 2014) resulted in 91 eligible SMS gateway service providers. The list of 91 services along with the links to their descriptions is available online (http://goo.gl/CcguZM). Evaluation of 91 services against 28 checks created a complex and challenging scenario for decision making as many of the services offered more or less the same functionality within the close price range. The first thing observed during the instantiation of ARISE in the case study was that it is not practical for evaluating the service specification or descriptions against requirements by following formal techniques, due to the huge number of available services and semantic heterogeneity in service descriptions by various service providers. However when compared to the service descriptions given in natural languages by the service provider, there are certain specific information that may not have been described. For example in the case study, some of the service descriptions were not clear about their payment mode and tax inclusion details. This information was retrieved later from the past users feedback and comments. When the past users' comments for all these services were retrieved with the help of web crawler and were further analysed it was found that there was a lot of irrelevant "noise" in those comments. They required further cleansing and parsing. The comments were analysed with "content analysis" technique and they were coded for the functionality for which the comment was reported. The comments were further categorized into positive and negative. Positive comments were all praising their respective service and not providing any useful information. However the negative comments were more informative. These comments were mainly about the quality and performance of the service such as delay time, reliability of the service to deliver the message. Without sentiment analysis scores and user comments, service 76 (Via SMS) appeared to be the best match. Whereas with all this available information, service 71 (Direct SMS) appeared to be optimally aligned with customer requirements, in terms of maximum coverage of the preferred requirements, as well as good reputation with the past users.

Table 1. Prioritized checklist from customer requirements [12]

Ri	Requirement check description
1	Service supports outgoing text messages in Australia
2	Service should not have any hardware or SIM requirements
3	Service should be highly reliable with 99.9% message delivery
...	....
26	Service shows notification of message delivery
27	Service shows message delivery failure notification
28	Service should provide schedule message delivery in case of holidays

The work in case study was conducted manually and took two weeks to complete the task. This highlighted the need for automation of some of the steps. The case study was helpful in identifying the requirements for the tool support for ARISE. In next section, we present the tool 'EVALUATOR' designed to support ARISE method and describe how the data from the case study was applied in the tool.

5   EVALUATOR – Tool for Service Selection

The aim of the tool is to assist the analysts with all the steps of ARISE while automating some parts of it, and provide a visual display of the quantitative and qualitative results at the same time. EVALUATOR requires the analyst to enter the set of requirements and service names (or descriptions); the tool uses an API for automatically calculating sentiment analysis and retrieving past user comments from internet. The tool would follow the steps of ARISE to convert the input into results which would be displayed in a graph with both numerical scores as well as textual comments. From the case study, it was obvious that ARISE method would require automation to make the tasks easier and less laborious for the analysts. The case study was helpful in identifying the requirements for the tool support for ARISE method and is required to provide following functionalities:

- Provide graphical interface to input requirement checklist and additional description and their associated weights for prioritization.
- Provide graphical interface to input service names, specification and additional data (online links, SLA, API description).
- Easy navigation to move back and forth in performing steps of ARISE
- Provide a grid interface for entering scores for all the services against every requirement in checklist.
- Provide facility to connect to online sentiment analysis sites and provide sentiment analysis data and popularity index of the services
- Provide facility for retrieving user comments and qualitative data of the services

- Show graphs and charts for calculating scores for all the services using the ARISE method
- Enable back and forth navigation and editing
- Maintain database for a specific project

Agile development methodology was followed for analysing, designing and implementing EVALUATOR. It was developed by using HTML5 and Javascript for client side scripting, and PHP for server side scripting. The database support was provided with MySQL. The interface for displaying results was supported by 'Highcharts' service API (http://www.highcharts.com/) and 'x10hosting' web server was used for uploading and deploying the EVALUATOR (http://evaluator.x10host.com/). In the following we provide the screen shots of the EVALUATOR tool which are in sequence to the steps of the ARISE method. The home page (Figure 2) is standard interface that provides options for login or creating new account. Also gives brief description of the method and a brief introduction to the tool. Help regarding the working of the tool is also available. Once the login is successful, a new project can be created. Also the previously stored projects can be retrieved form the database. The analysts can set values for requirements prioritization while they create the project. These are used for creating a range of MCDA weights that can be assigned to the requirements for calculating scores for the services. Next tab on screen is an input interface for requirements statements, weights and additional description or notes for further explanation (Figure 3). These weights are used for MCDA scoring and ranking as customer preferences for scoring services. The next interface (Figure 4) is for entering the names and URL of web services, and optionally the descriptions can be directly entered to database. Analysts can enter the scores for the services based on their granularity level to the requirements (Figure 5). This will automatically calculate score for individual requirement according to the customer preferences. The sentiment analysis is automatically calculated for every service by interacting with API from 'socialmention.com' by providing names of the services as input and getting both sentiments scores as well as past users' comments as output. This API is used for tracking any mention for the identified keywords in video, blogs, events, news, bookmarks, hashtags and even audio files. It categorizes the results into three types of sentiments i.e. positive, neutral and negative. It also gives values for "Passion" which is a measure of the likelihood that people who are talking about the product or service will do so repeatedly, and for "Reach" which is the measure of the range of influence of the product or service. "Strength" is the likelihood that the 'keyword' is discussed in social media within last 24 hours. These values are stored in database once they are generated. The analysts can update them later for new scores according to the time, as the sentiments can vary over the web each day, even each hour. The interface for results combines all the results into one compact display (Figure 6). Clicking on any resulting bar will show respective comments of users retrieved for that specific service (Figure 7).

6   Discussion

The main benefit of EVALUATOR was the automation of sentiment analysis and qualitative comments retrieval which took considerably long time when done manually. The visual aid also made it intuitively easier to see the trend among competing services by providing both qualitative and quantitative data in one compact view. Which service gets finally selected, is contextual and can vary in different projects. The main objective here was to observe if this additional source of information (user comments) were helping the analysts in overcoming the challenges of alignment in alignment for making informed decisions for service selection. The observations regarding some of the reasons that made alignment process challenging are discussed in following:

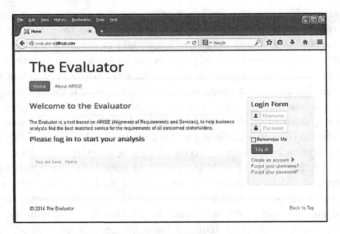

Fig. 2. EVALUATOR Home Page

- **Paradox of choice:** In the case study, there were a substantial number of related services to select from. MCDA scoring and ranking is a laborious and time consuming task when done manually. By automating this step of ARISE, it was helpful in reducing the laborious work in case where there is a huge sample of service specifications to evaluate.
- **Missing information in service specification:** In this case the comparison of requirements was done with service descriptions in natural language rather than with formal or technical documents. There were instances where the service providers were not giving fine details of the functionality e.g. in the case study, modes of payment for the service and tax related information. It was observed that mostly it is the non-functional requirements that are missing, whereas the basic functionality was described almost by all the service providers. Also every service provider had their own way of advertising their service giving rise to the diversity in semantic of the descriptions with which a specific requirement was to be compared. EVALUATOR helps in retrieving the user comments for further analysis and hence makes the data available for further assessment to find missing information.

Fig. 3. EVALUATOR Requirements Input Page

Fig. 4. EVALUATOR Service Input Page

Fig. 5. EVALUATOR Granularity Analysis Score Input

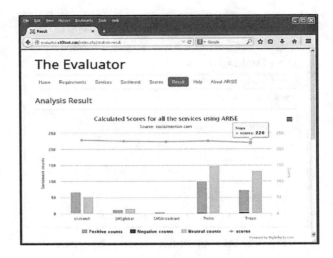

Fig. 6. EVALUATOR Analysis Results

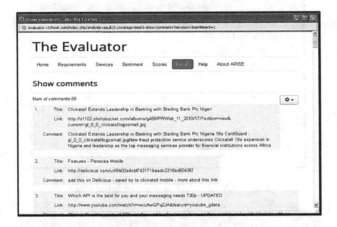

Fig. 6.   EVALUATOR Analysis Results – User Comments

- **Lack of user involvement:** Past user feedback and comments provide ways of involving them into the service selection decision process. The experiences of past users are a form of knowledge that can be brought into the alignment process for making more informed decisions. In ARISE method the past users of the services are involved through their feedback and reviews.

7   Conclusion and Future Work

The major contribution of the ARISE method is the involvement of the voice of past users with the help of sentiment analysis in the service selection process as well as utilising MCDA in the decision analysis. According to the results of the case study, it was observed that ARISE method does help in overcoming the challenges of service selection faced by the practitioners. It was also helpful in designing and developing of our supporting tool to reduce the information overload and assist the analysts to simulate the results for evaluating different options. In this paper we have presented a tool support for our previously proposed method of alignment of requirements and services (ARISE). The tool aims to assist the business analysts in making better decisions for service selection by providing qualitative as well as quantitative data regarding the sentiments of the past users of the service. The tool has been used in an observational case study for service selection. The results show that sentiment analysis helps in increasing information for business analysts, assists in making better informed decisions, and overcoming challenges of service selection.

We are currently experimenting with use of Natural Language Processing (NLP) tools to automate the step of MCDA and scoring, ranking and filtering of service specification. The idea of the approach is that a service description is likely to satisfy a requirement if it shares semantically related natural language content with it. Our future directions with EVALUATOR include integration of web crawler in EVALUATOR that can search for the service from online resources, and giving the tool ability to calculate sentiment scores for individual requirements rather than for a complete service.

References

1. Bano, M., Ikram, N.:. Issues and challenges of requirement engineering in service oriented software development. In: IEEE ICSEA, IEEE (2010)
2. Bano, M., et al.: What makes Service Oriented Requirements Engineering challenging? A qualitative study. IET Software 8(4), 154–160 (2014)
3. Papazoglou, M.P., et al.: Service-oriented computing: a research roadmap. International Journal of Cooperative Information Systems 17(02), 223–255 (2008)
4. Galster, M., Bucherer, E.: A business-goal-service-capability graph for the alignment of requirements and services. In: IEEE Congress on Services-Part I, IEEE (2008)
5. Gehlert, A., Bramsiepe, N., Pohl, K.: Goal-driven alignment of services and business requirements. In: International Workshop on Service-Oriented Computing: Consequences for Engineering Requirements SOCCER'08, IEEE (2008)
6. Bano, M., Ikram, N.: KM-SORE: knowledge management for service oriented requirements engineering. In: International Conference on Software Engineering Advances ICSEA (2011)
7. Bano, M., Zowghi, D.: User involvement in software development and system success: a systematic literature review. In: Proceedings of the 17th International Conference on Evaluation and Assessment in Software Engineering, ACM (2013)
8. Bano, M., Zowghi, D.: A systematic review on the relationship between user involvement and system success. Information and Software Technology 58, 148–169 (2015)
9. Seyff, N., Graf, F., Maiden, N.: Using mobile re tools to give end-users their own voice. In: International Requirements Engineering Conference (RE), IEEE (2010)
10. Liu, B.: Sentiment analysis and opinion mining. Synthesis Lectures on Human Language Technologies 5(1), 1–167 (2012)
11. Pang, B., Lee, L.: Opinion mining and sentiment analysis. Foundations and trends in information retrieval 2(1–2), 1–135 (2008)

12. Bano, M., Zowghi, D.: User voice and service selection: an empirical study. in Empirical Requirements Engineering (EmpiRE) at RE 2014, Sweden, IEEE (2014)
13. Chen, M., Liu, X.: Predicting popularity of online distributed applications: iTunes app store case analysis. In: Proceedings of the 2011 iConference, ACM (2011)
14. Fu, B., et al.: Why people hate your app: Making sense of user feedback in a mobile app store. In: Proceedings of the 19th ACM SIGKDD International Conference on Knowledge Discovery and Data Mining, ACM (2013)
15. Galvis Carreño, L.V., Winbladh, K.: Analysis of user comments: an approach for software requirements evolution. In: Proceedings of the 2013 International Conference on Software Engineering, IEEE Press (2013)
16. Seyff, N., Graf, F., Maiden, N.: End-user requirements blogging with iRequire. In: Proceedings of the 32nd ACM/IEEE International Conference on Software Engineering-Volume 2. ACM (2010)
17. Pagano, D., Maalej, W.: User feedback in the appstore: An empirical study. In: International Requirements Engineering Conference (RE), IEEE (2013)
18. Bano, M.: Aligning services and requirements with user feedback. In: International Requirements Engineering Conference (RE 2014), IEEE (2014)
19. Bano, M., Ikram, N.: Addressing the challenges of alignment of requirements and services: a vision for user-centered method. In: Zowghi, D., Jin, Z. (eds.) APRES 2014. CCIS, vol. 432, pp. 83–89. Springer, Heidelberg (2014)
20. Gu, Q., Lago, P.: Service identification methods: a systematic literature review. In: Di Nitto, E., Yahyapour, R. (eds.) ServiceWave 2010. LNCS, vol. 6481, pp. 37–50. Springer, Heidelberg (2010)
21. Huergo, R.S., et al.: A systematic survey of service identification methods. Service Oriented Computing and Applications, p. 1–21 (2014)
22. Bano Sahibzada, M., Zowghi, D.: Service oriented requirements engineering: practitioner's perspective. In: Ghose, A., Zhu, H., Yu, Q., Delis, A., Sheng, Q.Z., Perrin, O., Wang, J., Wang, Y. (eds.) ICSOC 2012. LNCS, vol. 7759, pp. 380–392. Springer, Heidelberg (2013)
23. Abelein, U., Paech, B.: Understanding the Influence of User Participation and Involvement on System Success–a Systematic Mapping Study. Journal of Empirical Software Engineering, p. 1–54 (2014)
24. Bano, M., Zowghi, D.: Users' involvement in requirements engineering and system success. In: IEEE International Workshop on Empirical Requirements Engineering (EmpiRE) (2013)
25. Damodaran, L.: User involvement in the systems design process-a practical guide for users. Behaviour & Information Technology 15(6), 363–377 (1996)
26. Hao, J., Li, S., Chen, Z.: Extracting Service Aspects from Web Reviews. In: Wang, F.L., Gong, Z., Luo, X., Lei, J. (eds.) Web Information Systems and Mining. LNCS, vol. 6318, pp. 320–327. Springer, Heidelberg (2010)
27. Harman, M., Jia, Y., Zhang, Y.: App store mining and analysis: MSR for app stores. In: Proceedings of the 9th IEEE Working Conference on Mining Software Repositories, IEEE (2012)
28. Pagano, D., Bruegge, B.: User involvement in software evolution practice: a case study. In: Proceedings of the 2013 International Conference On Software Engineering, IEEE Press (2013)
29. Vetschera, R.: Preference-based decision support in software engineering, In: Value-Based Software Engineering, p. 67-89. Springer (2006)

Author Index

Printed in the United States
By Bookmasters

Printed in the United States
by Bookmasters